Springer Tracts in Mechanical

Springer Tracts in Mechanical Engineering (STME) publishes the latest developments in Mechanical Engineering - quickly, informally and with high quality. The intent is to cover all the main branches of mechanical engineering, both theoretical and applied, including:

- Engineering Design
- Machinery and Machine Elements
- Mechanical structures and stress analysis
- Automotive Engineering
- Engine Technology
- Aerospace Technology and Astronautics
- Nanotechnology and Microengineering
- Control, Robotics, Mechatronics
- MEMS
- Theoretical and Applied Mechanics
- Dynamical Systems, Control
- Fluids mechanics
- Engineering Thermodynamics, Heat and Mass Transfer
- Manufacturing
- Precision engineering, Instrumentation, Measurement
- Materials Engineering
- Tribology and surface technology

Within the scopes of the series are monographs, professional books or graduate textbooks, edited volumes as well as outstanding PhD theses and books purposely devoted to support education in mechanical engineering at graduate and post-graduate levels.

More information about this series at http://www.springer.com/series/11693

Alessandro Freddi · Mario Salmon

Design Principles and Methodologies

From Conceptualization to First Prototyping with Examples and Case Studies

Springer

Alessandro Freddi
University of Bologna
Bologna, Italy

Mario Salmon
Bologna, Italy

ISSN 2195-9862 ISSN 2195-9870 (electronic)
Springer Tracts in Mechanical Engineering
ISBN 978-3-030-07018-2 ISBN 978-3-319-95342-7 (eBook)
https://doi.org/10.1007/978-3-319-95342-7

This Springer imprint is published by the registered company Springer Nature Switzerland AG
The registered company address is: Gewerbestrasse 11, 6330 Cham, Switzerland

To our Friend and Colleague Paolo Stanzani *(1936–2017), a great engineer and designer of wonderful cars, who would have been happy to collaborate on this book with his inventiveness and skill, whom a fatal disease prevented from fulfilling this dream.*

Foreword

A design principle can be defined as a general and scientific rule, which should have objective logic and pertinence, and might be commonly used in different design activities. Many researchers have probed whether scientific rules exist in engineering design. Design engineers and design theorists pursue the search for these rules. Of course, there are no definite answers to producing a well-designed product; however, there are tested principles which, if obeyed, increase the likelihood that a commercial product will be successful.

Good design is embedded in creativity, technical ability, communication at all levels, sensible management and the ability to forge all these attributes together with a multidisciplinary approach.

This book provides a reflection on the design of industrial products, seen as a pivotal and distinctive activity of the engineer. As an example of interdisciplinary activity, design needs the contribution of different cultures, from technical to human sciences (statistics, experimental analysis, cognitive sciences, aesthetics, logic, philosophy, history, etc.).

Moreover, the book offers a thorough introduction to several design methods, their differences and similarities and outlines the role of design standards and guidelines, such as understanding customer needs, examining the notions of invention and innovation, dealing with the robustness of the concept and focusing on the design process through analysis of case studies.

Addressed to engineering students, as well as professional engineers and design practitioners, it collects and merges the visions of two researchers characterized by different long-term professional and academic experiences, who, however, share a common academic education as both are mechanical engineers. As always happens when a book is written collaboratively, a sharing and a complementarity of thought are necessary, which should be corroborated, as is the case with the authors, by friendship and mutual respect, which are important prerequisites for the success of the venture.

Bologna, Italy
May 2018

Ezio Mesini
Dean of the School of Engineering
and Architecture
University of Bologna

Preface

The purpose of this book is to give a preliminary learning guidance on concepts which serve as the basis of design instruction. This is a primer to guide the aspiring designer in the gaining of insights into design thinking. It represents the synthesis of teaching experience, in the hope that it can be useful to different cases.

Designing is a multidisciplinary activity and cannot be totally entrusted to acquiring industrial experience. It requires a methodical path of learning. This does not detract from the fact that the practical dynamics of design activity are often different and complementary to what is being studied in the books devoted to design apprenticeship. For this reason, the book presents the synthesis of two points of view of two different authors: the first one academic and the second from the industrial design practice. The authors have different professional experience, and the book collects and tries to combine two backgrounds, without forcing them into a single thought. They hope that readers will be able to draw their own, personal design view from the book.

Every design method requires reflection, that is, a way to review the experience: nobody can acquire expertise in just one step, but any apprenticeship requires stratification of knowledge. Not only: as is reported in Chap. 9, following Dante Alighieri's reasoning, in any human activity there is always a gap between theory and any actual implementation. The authors think that design is an open system in which the young scholars that wish to cultivate it must find their own path freely.

The book is limited to conceptual design up to prototype building, and it does not deal with industrialization, manufacturing and project management themes. As regards the meaning of the word "concept", this is intended as a clear, detailed description of the attributes and benefits of a new product that addresses the needs of the targeted customers. So conceptual design is a phase of clarification and definition up to a first realization.

As regards the use of mathematics, formalization is often an obstacle to general understanding and leads to thinking that design is only the task of specialized actors, an idea that is contrary to the multidisciplinary nature and to inventiveness. However, it was not possible to avoid mathematical presentation for some topics; it was necessary for them to refer to probabilistic and statistical models. These topics, after the brief introduction of this book, can be studied in more depth through specialized sources, cited in the bibliography.

A final point: according to a method often applied in this book, many definitions of concepts, words, methods, tools, etc., were taken directly from the technical literature, appropriately quoted. The technical literature on design is extensive but not always consistent and uniform. The best-judged definitions are given, sometimes taken from the Web.

The chapters devoted to an industrial experience demonstrate that often, in practice, it is necessary to follow other design paths, based on emerging priority needs, e.g. on new problems that suddenly arise, rather than to start from the beginning of the whole design process. For this reason, the book wants to establish a reference to the main concepts but rejects any form of normativism and prescriptivism.

Bologna, Italy
Summer 2018

Alessandro Freddi
Mario Salmon

Acknowledgements

The authors would like to thank Matteo Lena, CEO at SASIB S.p.A. (a company of the Coesia Group), for permitting the use of manuals and product documentation for the description of the ALFA automatic packaging machine.

A special thanks to Alberto Pratelli for providing the first ALFA prototype style drawings and other industrial design ideas developed with one of the authors. This material has allowed a reconstruction of some of the conceptual design phases which, without these notes, would end in oblivion. The authors would like to mention also the contribution of Umberto Meneghetti to the design of the ALFA machine.

Thanks must also go to several professors of the School of Engineering of University of Bologna for their feedback during the presentation of some of the main topics of the book. Among them, Ezio Mesini and Domenico Mirri invited the authors several times to present their educational proposals to the School of Engineering.

With respect to the design of prototype, special thanks go to Sergio Curioni, co-developer of many laboratory devices, and Vincenzo Dal Re, Gianni Caligiana, Giorgio Olmi and other colleagues for the numerous collaborations and fruitful discussions (see the citations of related papers in the bibliography).

A debt of gratitude also goes to Alessandro Birolini, a distinguished scientist in the field of reliability, who made a friendly revision of a crucial part of Chap. 5, and to an old-time friend, Giulio Cesare Barozzi, for his mathematical insights. Thanks also to Roberto Scazzieri for the enriching discussions about innovation in economic processes.

Finally, family members such as Ivana and Maria Rosa deserve special thanks for the time they gave up during the preparation of this book.

Contents

About the Authors

Alessandro Freddi was Professor of Experimental Stress Analysis and Principles and Methodologies at the School of Engineering of University of Bologna. Over the years, he has provided consultancy for various Italian industries. He has presented some of the results of his research at the Cambridge Engineering Design Centre (EDC) and at Bundesanstalt für Materialforschung und -prüfung (BAM) in Berlin. He is Professor Emeritus of the University of Bologna, Life Member of Clare Hall Cambridge, Benedictine Academician of the Academy of Sciences of the Institute of Bologna, Member of the Hungarian Engineering Academy and Honorary Professor of Wroclaw University of Science and Technology (Poland).

Mario Salmon graduated in Mechanical Engineering from University of Bologna. He was employed mainly in the industrial sector at the Italian Generating Board (ENEL), CERN Geneva, and Olivetti Ivrea, where he was responsible for the design of a new machine tool, SIGMA. He was then R&D Director at SASIB Bologna Packaging Machinery Division, where he developed the Alfa Packaging Machine together with Elio Pagella. He also serves as a contract Professor for Robotics and Information Technology Application at Milan Polytechnic and at the University of Bologna. Since 1984, he has also been active as a freelance industrial consultant.

Introduction

The book is a primer intended as a first approach of principles and methods that form the basis of engineering design. As regards the importance of engineering design, the authors share the thinking of Herbert A. Simon, [1, 2].

> ... Schools of engineering, as well as schools of architecture, business, education, law, and medicine, are all centrally concerned with the process of design. In view of the key role of design in professional activity, it is ironic that in this century the natural sciences almost drove the sciences of the artificial from professional school curricula, a development that peaked about two or three decades after the Second World War. Engineering schools gradually became schools of physics and mathematics; medical schools became schools of biological science; business schools became schools of finite mathematics. [...]
>
> Such a universal phenomenon must have had a basic cause. It did have a very obvious one. As professional schools, including the independent engineering schools, were more and more absorbed into the general culture of the university, they hankered after academic respectability. In terms of the prevailing norms, academic respectability calls for subject matter that is intellectually tough, analytic, formalizable, and teachable. In the past much, if not most, of what we knew about design and about the artificial sciences was intellectually soft, intuitive, informal, and cook-booky. [...]
>
> The damage to professional competence caused by the loss of design from professional curricula gradually gained recognition in engineering and medicine and to a lesser extent in business. [...]
>
> What is called for is not a departure from the fundamental but an inclusion in the curriculum of the fundamental in engineering along with the fundamental in natural science. That was not possible [once]; but it is possible today. [...].

All this prompted the specific adopted approach, aimed at training new generations of engineers involved in design activity, by offering study programmes in product and service design, placing the emphasis on creativity, technology and design methodologies with a concern for human values and society's needs. The presentation combines the exposure of the methodological aspects of design (such as ISO standards, VDI guidelines, [3] and others such as Pahl and Beitz's approach) and the reference to general descriptive principles derived from cognitive processes, [4].

The book is arranged in three parts, articulated in the following chapters:

- Chapter 1 introduces some epistemology concepts of *Engineering and Industrial Design* and gives some criteria for classifying and evaluating them. It discusses the difference and quotes a theory that distinguishes the working process of both [5, 6].
  Examples are shown for some technical products of an Italian region, rich in design tradition of cars, engines, motorbikes and packaging machinery.
  Part I focuses on the analysis of principles and methods of design.
- Chapter 2 presents some design methods and the role of design standards and guidelines.
- Chapter 3 is devoted to the understanding of customer needs and introduces a tool for transforming confused requirements into clear technical and measurable data.
  This topic is today one of the most controversial points of view that addresses speculation on users' inferences and on the models that respect users' active engagement with designed products [7].
- Chapter 4 introduces the concepts of *invention and innovation* as necessary clarification in understanding the design tasks and the importance of innovation strategies in the industrial environment.
  Part II is an introductory presentation of two topics that need mathematical formalism. It consists of two chapters:
- Chapter 5 deals with *Safety Assessment*, i.e. with Design to reduce (eliminate) failures.
- Chapter 6 deals with *Design of Experiments.*
- Chapter 7 deals with *Introduction to the Taguchi Method*, i.e. with robustness concept.
  Part III deals with applications of the design principles to industrial products.
- Chapter 8 covers One-off Product Design, i.e. the design and building of single laboratory equipment, following a simplified "evolutive" point of view.[1]
  These case studies, even limited to prototype phase, are suitable to illustrate some systematic design tools which can be extended to a general use (e.g. functional analysis, affinity diagrams, relationship and interrelationship matrices, morphological matrices, value analysis).
- Chapter 9 focuses on *the conceptual design of a packaging machine.* Its design is a paradigmatic example of inventions and innovative products of a special geographical area.
- Chapter 10 deals with *the embodiment design of the previous packaging machine.*

This book wants to provide a first approach to the numerous aspects of design, in order to favour competencies in the professional and academic domains of apprenticeship of young designers; each chapter of the book can then become a reason for in-depth study of subjects in specialized books. This book therefore

[1] New designs have roots in previous designs.

requires subsequent study to explore all the specialist aspects of design science, relatively to:

- Invention [8–14],
- Other systematic methods [15, 16],
- Design management [17],
- Reliability [16, 18, 19],
- Design for assembly [20].

All these recent books widely testify to the removal of design from the practical realm based only on apprenticeship and experience and place it in a scientific context.

Finally, the Appendix offers a brief *Packaging Valley History*. The Packaging Valley is an area of a few hundred square kilometres: its boundaries usually coincide with some provinces of Emilia-Romagna administrative region in Northeast Italy. This human and technical environment has strongly influenced the development of design. The Appendix tries to briefly analyse the reasons for such an exceptional concentration of similar industries.

References

1. Simon HA (1996) The sciences of the artificial, vol 1, 3rd edn. MIT Press, USA. ISBN10 | asin : 0262193744 print ISBN13: 9780262193740 ebook ISBN13: 9780585360102
2. Christine Keitel-Kreidt KR (2012) Learning from computers: mathematics education and technology. Springer Science and Business Media, the NATO Advanced Research Workshop on Mathematics Education and Technology, Villard-de-Lans, France, May 6–11, 1993
3. VDID (2016) Verband Deutscher Industrie Designer e.V. http://www.vdid.de/
4. Lidwell W, Holden K, Butler J (2003) Universal principles of design. Rockport Publication (I), Beverly, USA
5. Liu S, Boyle IM (2008) Engineering design: perspectives, challenges and recent advances. J Eng Des 20(1):7–19
6. McMahon CA (2012) Reflections on diversity in design research. J Eng Des 23(8):563–576
7. Crilly N (2011) Do users know what designer are up to? product experiences and persuasive intentions. Int J Des 5(3):1–15
8. French M (1994) Invention and evolution: design in nature and engineering, vol 1, 2nd edn. Cambridge University Press, Cambridge
9. Cross N (2008) Engineering design methods: strategies for product design, vol 1, 4th edn. Wiley, Hoboken, ISBN: 978-0-470-51926-4
10. Dym CL (2012) Engineering design : representation and reasoning, 2nd edn. Cambridge University Press, Cambridge
11. Dieter GE (2013) Engineering design, 5th edn. McGraw-Hill, New York
12. Myrup Andreasen M, Hansen C, Cash P (2015) Conceptual design. Interpretations, mindset and models, vol 1, 1st edn. Springer, Berlin
13. Bayazit N (2004) Investigating design: a review of forty years of design research. Des Issues 20(1)
14. Wikipedia (2017) Design methods. https://en.wikipedia.org/wiki/Design_methods

15. Pugh S (1996) Creating innovative products using total design: the living legacy of Stuart Pugh. Addison-Wesley, Reading
16. Pahl G, Beitz W, Feldhusen J, Grote KH (2007) Engineering design: a systematic approach, 3rd edn. Springer, London
17. Hales C, Gooch S (2004) Managing engineering design, 2nd edn. Springer, Berlin
18. Birolini A (2017) Reliability engineering, 8th edn. Springer, Berlin
19. Ulrich KT, Eppinger S (2008) Product design and development, 5th edn. Mc-Graw Hill, ISBN: 978-0073404776
20. Ullman D (1992–2009) The mechanical design process, 4th edn. McGraw-Hill, New York, ISBN 13: 9780071267960

Part I
Methods

This part of the book provides an overview of the methods that deal specifically with product and services design. Design is always a challenge for the future but with the support of past experience. Design draws its knowledge from experience and from physical sciences for addressing the needs of human beings that are always the final recipients. Even for the most abstract projects, the role of design must have a positive result for humanity, the environment, and the universe. Otherwise, the science of building artifacts would become a curse for humanity. For these reasons, professional associations have upgraded the designer's role over the years to the point of giving great importance not just to technical values.

From the point of view of the philosophy of science, design is largely based on technical sciences as well as on human sciences and is at the crossroads of reductionist and holistic thinking.

A method derives from the personal experience of the designer, but it aspires to reach the statute of a universal approach. Basically, methods can broadly be categorized according to two lines of thinking. To the first line belong the methods that are based on "evolutionary" thought, according to the statement that each new product/service is, in fact, the evolution of an existing one. They are based on the observation of strengths and weaknesses of similar products on the market, already tested by the users' experience.

To the second line belong the "systematic" methods which, on the basis of a reductionist logic, consider design a set of abstract functional relations with which principles of realistic solutions may be associated. In this way, at least theoretically, a new design problem can be solved without dependencies on previous choices.

Both approaches have some truth value, and the current presentation leaves the reader freedom to build a personal opinion about them.

Chapter 1
Engineering Design and Industrial Design

Abstract Engineering designers and industrial designers are the main actors called to collaborate, within their jurisdiction, in the design of products/service through creative choices that define functions, structures and forms together with manufacturing processes. In spite of different cultures and practical approaches of the two categories of professionals, design accomplishes a common goal: development of new products and services. Each design process runs from a conceptualization phase and moves towards an embodiment phase, trying to meet the customers' real needs, and at the same time to satisfy business requirements, in respect of health-, safety- and environmental constraints [32]. Obviously, it is necessary to distinguish between the end user and the industrial customer that have requirements that must be met differently.

1.1 Epistemology of Design

If the creation of artifacts is one of the main roles of engineering, design is its central activity because it's just the design that distinguishes engineering from the pure sciences [15]. On the other hand, we must frankly admit that also industrial designers consider design as their distinctive profession.

In order to understand the different perspectives of engineering and industrial designers regarding product/service design, the first analysis to be carried out is comparing the epistemic content of the two professions.

1.1.1 Multidimensional Aspects of Design Thinking

After this premise, it may be useful to revisit an idea [36] that goes back to the 1st century B.C., having been laid down by *Vitruvius*, the famous Roman architect and engineer, in his book [43, 44]. He first clarified the multidimensionality and the multi-facetted values of design that even modern culture asserts. He identified the following three general categories that are recognizable in design: *Utilitas, Firmitas and Venustas*.

A. Freddi and M. Salmon, *Design Principles and Methodologies*, Springer Tracts in Mechanical Engineering, https://doi.org/10.1007/978-3-319-95342-7_1

Concordances and relationships of these old categories with some of the modern ones are the following:

1. Utilitas is related to Function.
2. Firmitas is related to Structure.
3. Venustas is related to Aesthetics.

Utilitas is the first requirement of a design. *Firmitas* refers to the correct organization of its parts and also to their strength and robustness. *venustas* is the value that is "beyond the functionality" and "beyond the structure" related with aesthetics, form, appearance etc.[1] With the help of the previous analysis, it is thus possible to identify basic principles to which we should refer in judging existing designs and in carrying out new ones.

1.1.2 Reductionism and Holism

Extending the interpretation of [26] (with the quotation of Scherwin, Henry and Hisherwood) and [18], we may assume that the epistemology of engineering design is related to:

- Technical knowledge.
- Practical knowledge.

This multidimensional nature of design suggests an epistemic interpretation, as interaction of *reductionism and holism.*

We can schematically say that reductionism is a type of knowledge of physical and technical sciences (basic and engineering sciences), while holism is a type of knowledge of form design. Their definitions are taken from a dictionary:

> Reductionism = The practice of analyzing and describing a complex phenomenon in terms of its simple or fundamental constituents: reductionism is a typical methodology used in physical sciences to attempt to provide explanation in terms of ever smaller entities, and on the assumption that an optimal choice exists for the solution of each individual function.
> Holism = the theory that parts of a whole are in intimate interconnection, such that they cannot exist independently of the whole, or cannot be understood without reference to the whole, which is thus regarded as greater than the sum of its parts.

While reductionism advocates the principle of breaking down systems into constituent parts in order to be reconstructed into a single whole solution at a later stage, holism attempts to capture the essence of systematic behaviour rather than that of the individual parts of a system [26], Fig. 1.1.

---

[1] It is also peculiar that Vitruvius' definitions acquire a special meaning if they are related to communicative contents of a design. Utilitas, firmitas and venustas recall three linguistic categories:

1. *pragmatics*: that deals with function, i.e. the study of what words mean in particular situations (what their function is).
2. *syntax*: that deals with the relations between expressions, i.e. with their structure.
3. *semantics*: that deals with the relations between signs and what they denote. Semantics classifies and examines changes in meaning and form.

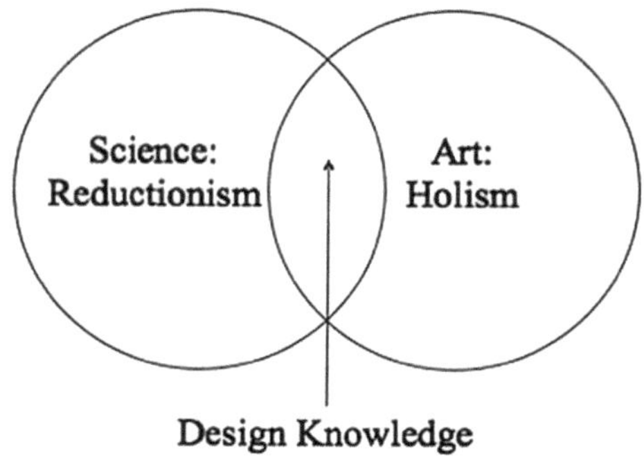

Fig. 1.1 Epistemology of design: combination of reductionism and holism

1.2 Definitions

In order to understand the common target (if it exists) and the differences between Engineering and Industrial Design, the starting point is the definition comparison of the different professional associations. In order to do this, we compare the contents of a curriculum on engineering design education (established by the Accreditation Board for Engineering and Technology, (ABET) with some other definitions coming from the International Council of Societies of Industrial Design (ICSID), and the Industrial Designers Society of America (IDSA).

1.2.1 How Engineering Design Is Defined

The point of view of engineering design is well summarized in a definition by ABET,[2] [1–3]:

> Design is the process of devising a system, component, or process to meet desired needs. It is a decision making process (often iterative), in which the basic sciences, mathematics, and engineering sciences are applied to convert resources optimally to meet a stated objective. Among the fundamental elements of the design process are: the establishment of objectives and criteria, synthesis, analysis, construction, testing and evaluation.
> The design component of a curriculum must include some of the following features: the establishment of objectives and criteria, synthesis, analysis, construction, testing and evaluation, development of student creativity, use of open ended problems, development and use of design methodology, formation of design problem statements and specifications, consideration of alternative solutions, feasibility considerations, and detailed system descriptions. Furthermore, it is essential to include a variety of realistic constraints such us economic factors, safety, reliability, aesthetics, ethics, and social impact. *See also the analysis in* [6].

---

[2](415 North Charles Street, Baltimore, MD 21201).

1.2.2 How Industrial Design Is Defined

Conversely, the definition of Industrial Design has changed over the years. This is probably due to a progressive awareness of its technical and social role. At the beginning of 1959, the profession of an Industrial Designer was defined in a very limited and finite way [22]:

> An industrial designer is one who is qualified by training technical knowledge, experience and visual sensibility to determine the materials, mechanisms, shape, color, surface finishes and decoration of objects which are reproduced in quantity by industrial processes. The industrial designer may, at different times, be concerned with all or only some of these aspects of an industrially produced object.

In 1969, the previous definition was enlarged, tha nks to the contribution of some important scholars [9–11]:

> Industrial design is a creative activity whose aims is to determine the formal qualities of objects produced by industry. These formal qualities are not only the external features but are principally those structural and functional relationships which convert a system to a coherent unity both from the point of view of the producer and the user. Industrial design extends to embrace all the aspects of human environment, which are conditioned by industrial production.

Recently, at the 29th General Assembly of ICSID in Gwangju, South Korea in 2015, the Professional Practice Committee unveiled an updated definition of industrial design [23]:

> Industrial Design is a strategic problem-solving process that drives innovation, builds business success and leads to a better quality of life through innovative products, systems, services and experiences. Industrial Design bridges the gap between what is and whats possible. It is a trans-disciplinary profession that harnesses creativity to resolve problems and co-create solutions with the intent of making a product, system, service, experience or a business, better. At its heart, Industrial Design provides a more optimistic way of looking at the future by re-framing problems as opportunities. It links innovation, technology, research, business and customers to provide new value and competitive advantage across economic, social and environmental spheres.
>
> Industrial Designers place the human in the centre of the process. They acquire a deep understanding of user needs through empathy and apply a pragmatic, user centric problem solving process to design products, systems, services and experiences. They are strategic stakeholders in the innovation process and are uniquely positioned to bridge varied professional disciplines and business interests. They value the economic, social and environmental impact of their work and their contribution towards co-creating a better quality of life.

In 2009, the Industrial Designers Society of America (IDSA) gave the following, further definition of the Designer's task, highlighting the community contribution [25]:

> Industrial Design (ID) is the professional service of creating products and systems that optimize function, value and appearance for the mutual benefit of user and manufacturer. Industrial designers develop products through collection analysis and synthesis of data guided by the special requirements of their client and manufacturer, with clear and concise recommendations and through drawings, models and descriptions. Industrial designers improve as

> well as create, and they often work within multidisciplinary groups that include management, marketing, engineering and manufacturing specialists.

Other Associations are present in different countries to promote the designer's professional activity, e.g. [4, 5, 7, 12–14, 20, 21, 24, 42].

1.3 Two Different Approaches

It is also clear that, in product (and service) development, other actors are on stage [29], but engineering and industrial design still have a leading role.

> [...] Manufacturers who are properly concerned with product quality will engage teams of engineering and industrial design people to create artifacts that work, look nice, and are easy to produce. Thus engineering design and industrial design are viewed as the major elements of product design that take effort in the practical design activity.

The previous comparison is a good starting point for the analysis of differences and similarities, but it is not sufficient to diversify the tasks of these two professional categories in production practice. An observational analysis, highlights the following distinctive features:

- Engineering designers include in the product categories not only single objects but also industrial machinery, typical of the secondary sector.[3]
- Engineering designers consider an innovation of a production method important as a new product.
- Industrial designers are interested not only in "objects which are reproduced in quantity by industrial processes" but also in crafted products.
- Industrial designers are particularly sensitive to fashion and to the tastes of end users.

The example of Fig. 1.2 shows the design of a cover for a cooling circuit in a car engine. This problem was technically solved by engineers, but was drawn together by an engineer and an industrial designer.

A theoretical controversy on design role could be summarized in the following points. The typical theoretical position of engineering:

- *Function first, afterwards form*: Form follows Function [28]. Two radical interpretations derived from this theoretical position:

  1. A descriptive interpretation: beauty results from purity of function.
  2. A prescriptive interpretation: a designer must follow functional considerations since aesthetic values are of minor relevance.
     A critique to this position is summed up in this statement: *No form derives necessarily from its function.*

---

[3]Secondary sector is the portion of an economy that includes manufacturers of finished goods and products from raw materials, and of machines that are not final products but are designed to produce finished products for end users (e.g. machines for packaging goods) [39, 40].

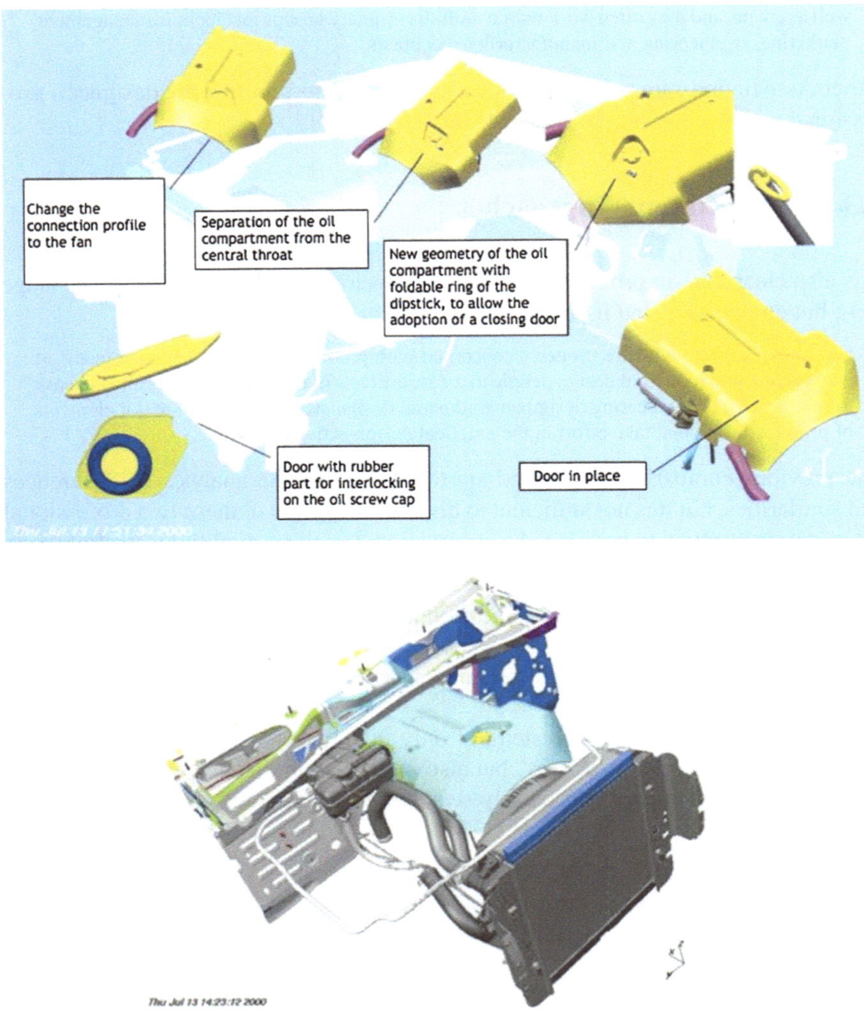

Fig. 1.2 Example of a design of an engine cover (SSPDI a Course on Design at a Summer school of Bologna University: F. Belletti, Tutor P. Stanzani): evolution of the form from the original drawing. Modified from [19]

- A totally opposite position, typical of art designers, is summed up in this contrasting statement: *Form first, afterwards function*.
  This statement derives from the observation that form is often a creativity sign that does not respect the context, since form itself creates new uses and contexts. The consequent risk is to treat only the formal values with the extreme position of so-called *styling*.

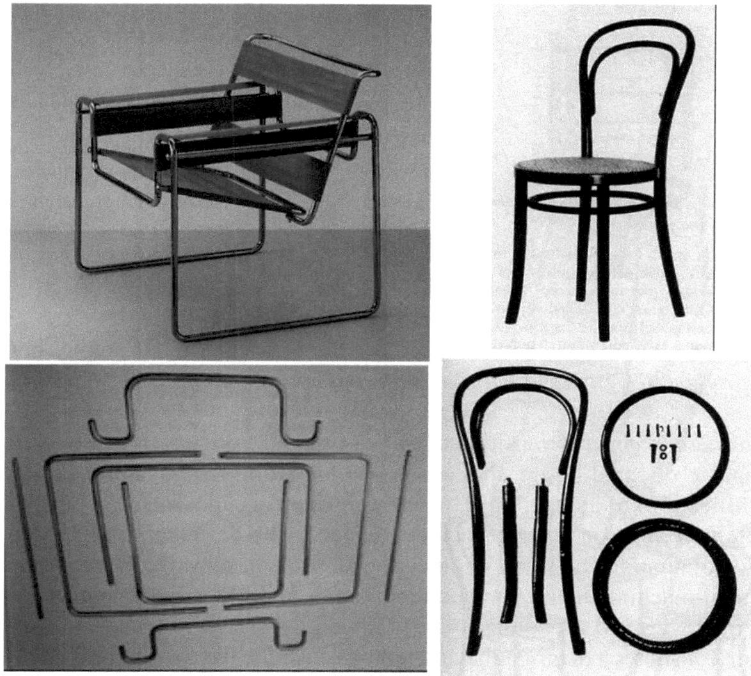

Fig. 1.3 Comparison between the semplicity of the Arm-chair by M. Breuer (1925) and the chair Thonet N.14 (1860), typical examples of good industrialized designs (design for assembly). Modified from [38]

For a discussion on this point and the relative bibliography see [28].

Before explaining a general theory about the difference between engineering and industrial designs, we briefly discuss another point.

A substantial difference is present between the design of objects intended for industrial production (even in small quantities) and objects produced with a strong contribution of craftsmanship, i.e. manual activity.

Two common examples clarify this point due to a reflection of [38]. The first example is the famous arm-chair designed by Breuer. It is a product conceived for industrial production that utilizes, as a structural frame, steel tubes assembled by screws (Fig. 1.3). The frame is produced industrially.[4]

The second example is an arm-chair designed by Mies van der Rohe, presented again in [38]. It is a crafted artifact, not really an industrial product, since it needs slow, time waster manual intervention of craftsmen for producing, adjusting and

[4]Even in the automotive and motorcycle industries the tube frame can be a structural solution, periodically revived in modern vehicles. This solution applied a principle of simplicity to the production technology.

Table 1.1 Comparison between a technical and a humanistic oriented design approach

Technically oriented	Customer's satisfaction oriented
Minimum production cost	Design for cost
Minimum space requirement	Design for reliability
Minimum weight	Design for assembly
Minimum losses	Design for robustness
Optimum handling	Design for ergonomics
	Design for X

smoothing the intersection between the legs. Summarizing, the example shows that *not all product design is industrial design and not all mass products are industrial design.*

In any case, a general transition must be observed from a technically oriented to a humanistic oriented design that arises by comparing the general technical rules of late fifties [33] to today's rules. Modern design concepts focus on customer satisfaction more than on abstract technical values. The designer's profession shows convergent intentions between Engineering Designers, sensitive to human sciences and to man-machine interaction and ergonomics and Industrial Designers, open to new technologies and economics trends (Table 1.1).

Figure 1.4 shows a diagram that attempts to describe the two areas of engineering and industrial designers, showing an area of common interest. The fundamental common items of a product/service design in the first column will be seen in detail later [33]. They are: Safety, (with all the regulations and standards established by the laws and specialized safety rules), Ergonomics, Production, Control, Assembly, Transport, Operation, Maintenance, Time and Costs, Recycling. The typical fields of action of Engineering and Industrial Design are compared in the second and third columns.

1.3.1 Inside-Out and Outside-In Working Process

An original point of view about the different approach of engineering and industrial design is described in [27]. See also [30, 31]. The theory refers more to the practice of working process than to a real epistemic difference:

1. Engineering design is developed *inside-out*, focused on function and structure (structural design).
2. Industrial design operates *outside-in*, by starting from the outer form then proceeding to the design of the internal parts.

If a conflict between the two points of view is created and not solved, compromises may arise, see example in Fig. 1.5.

Correspondingly, following this theory, the explained different approaches are the source of two different types of functions:

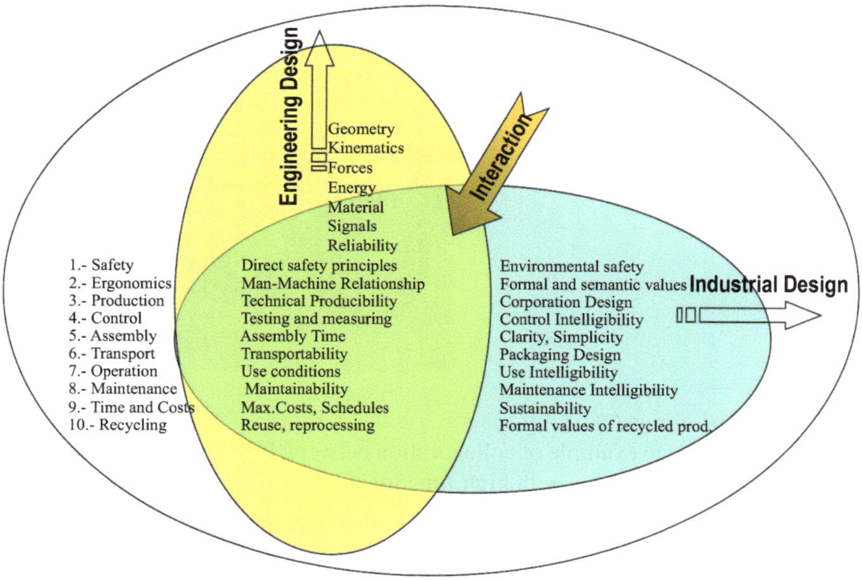

Fig. 1.4 Two major parts in product design with the area of engineering and industrial design interaction. Modified from [19]

Fig. 1.5 Example of a protuberance due probably, to an unresolved conflict between industrial design and engineering design

1. Product-working function.
2. Human-using function.

After this theoretical diversification, let us show, on the contrary, examples of good design interactions, with special reference to a geographical environment rich in product design tradition.

We will distinguish cases of product design for final users from cases of product design for industrial customers.

1.3.1.1 Interaction for Design of Products for Final Users

Motorcycles are typical examples where *function, structure and form* have equal values for advanced users.

Fig. 1.6 Example of a product with functional and semantic values (Ducati, Bologna) Designer: Miguel Galluzzi

Example 1.1 (*Motorcycle*)

Figure 1.6 shows an example of collaboration between Engineering and Industrial Design, resulting in solutions with high functional and semantic values.

The framework consists of struts joined by welding. The visible structure (without any cover, called "naked") communicates ideas of stiffness and resistance and is essential for successful market reception [28]. The functional innovation of engineering design is likewise shown in Fig. 1.7, where the technical principle of desmodromic valve command is adopted.[5]

The engineering designer was quoted to say to support innovation [45]:

> The specific purpose of the desmodromic system is to force the valves to comply with the timing diagram as consistently as possible. In this way, any lost energy is negligible, the performance curves are more uniform and dependability is better.[6]

Example 1.2 (*Sport Cars*)

Another example of innovation due to good interaction between engineering and industrial design was given, several years ago (1971 ÷ 1990), by the solution principles adopted on a Lamborghini car (S. Agata-Bologna IT), when the engine was mounted in a reverse arrangement inside the body of the Countach LP400 model. This choice favored the weight distribution and the provision of an all wheel drive system. Furthermore, this design pioneered the wedge-shape of the car. It also popularized the *cabin-forward* design concept, which pushes the passenger compartment forward to accommodate a larger engine at the back. It was engineered by Paolo

[5]It is a reciprocating engine valve that is positively closed by a cam and leverage system, instead of a spring. Force is directly applied to the valve stem, without the interposition of a spring. Twin-cylinder engines were developed and produced according to this solution by Azzariti, an Italian manufacturer from 1933 to 1934. After application of this solution principle by Mercedes-Benz on Formula One racing cars in 1954–1955, and on Mercedes-Benz 300SLR sports racing cars, in 1956, Fabio Taglioni applied it in Ducati motorbike engines. A desmodromic valve actuation has been applied to top-of-the-range production Ducati motorcycles since 1968.

[6]However, it is necessary to observe that the solution requires very careful control of the clearance and consequently very complex and expensive processing cycles and adjustment.

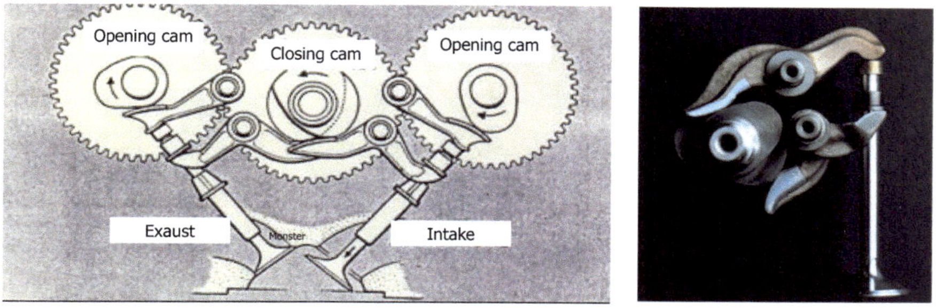

Fig. 1.7 Desmodromic valve command. Engineering Chef-Designer: F. Taglioni

Fig. 1.8 Design of Lamborghini Countach car's body due to M. Gandini (on the left) and of a modern model (on the right)

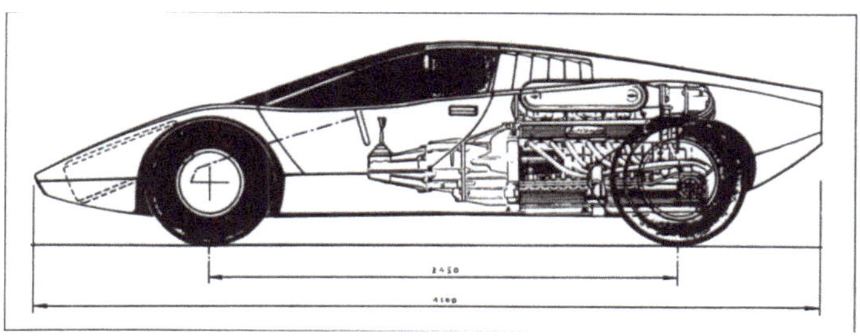

Fig. 1.9 Lamborghini Countach car's reverse engine arrangement by Chef Engineer P. Stanzani

Stanzani [16], with the line designed by Marcello Gandini, Fig. 1.9. This is an example of a close cooperation between engineering and industrial design. The basic solution has continued to be adopted even to the advent of new technologies of car design, Fig. 1.8.

1.3.1.2 Design of Industrial Machinery

In this case too, where the functional content prevails on the line design, industrial design had the role to improve not functional properties, but other characteristics

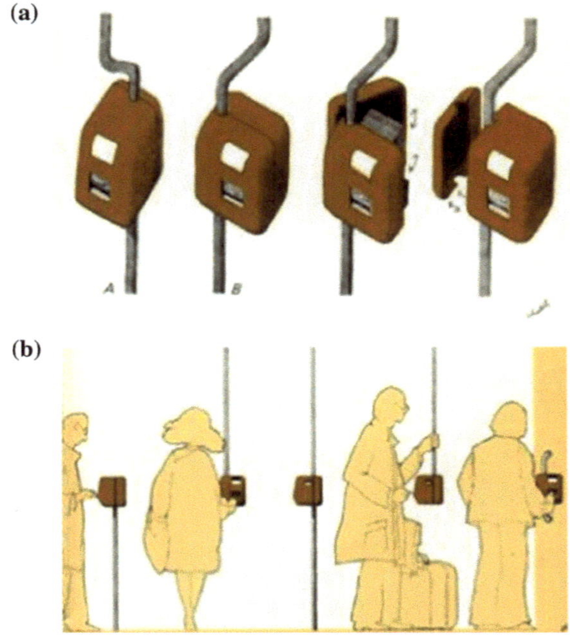

Fig. 1.10 Ticket validating machine. Industrial Designer: Alberto Pratelli. Engineering Designer: Mario Salmon [35]

such as ergonomics for improvement of the user's interface [28], for maximizing usability. While the design of computers, mobile phones and other personal information systems has received great attention on the part of designers, a potential exists for industrial machinery due to the variability of the subjects for which interfaces are intended in an industrial environment. Design- related requirements are viewed as a priority safety consideration not only for direct operators, but also for other control and maintenance officers and for all those who interact with the machine, (for maintenance, safety control, etc. [8, 17, 41].

Example 1.3 (*Ticket Machine*)

The example deals with the design of a ticket validating machine, Fig. 1.10. This is an automatic machine designed for bus users etc. [35] The ergonomic requirements are on:

- *Accessibility*, i.e. the machine must be designed to be usable by as many persons as possible, with inclusion of people with diverse abilities. Accessible design must be of easy perception, good operability, simplicity and forgiveness.
- *Affordance*, i.e. the machine must be designed so that its physical characteristics influence its function. It must be affordable to mount in different positions and to be used by persons of different cultures and intelligence quotients. The design of the buttons and of the slit for the introduction of the tickets must follow the same principle, Fig. 1.11.[7]

---

[7]The introduction of the icons on the computer desk are a typical example of application of this concept.

(a) *Detail of the slit* **(b)** *Detail of the clock*

Fig. 1.11 Details of external parts of the machine

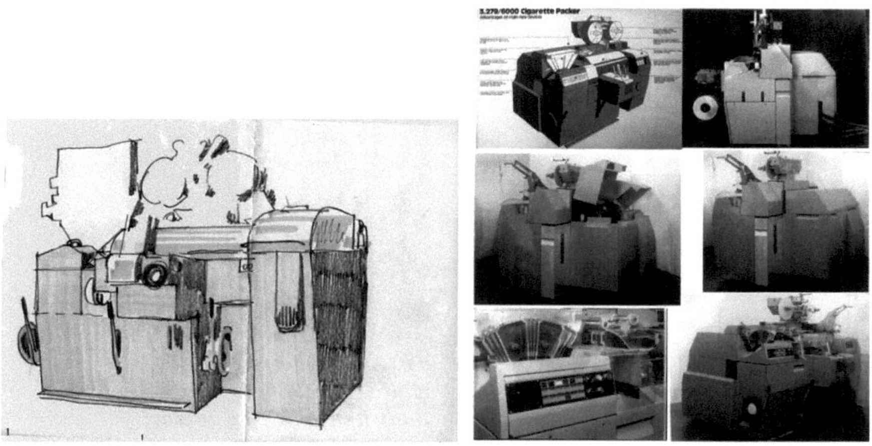

Fig. 1.12 Sasib packaging machinery bodies (Designer: A. Pratelli)

1.3.1.3 Industrial Design of Packaging Machinery

The Design of a Packaging machine will be presented in detail in a specific Chapter. Here the design of an automatic machine is shown to demonstrate that, also in this extreme context of pure industrial machinery, a contribution of industrial design is also desirable, for the quality of life of operators.

Example 1.4 (*Packaging Machinery*)

Figure 1.12 shows examples of different models of packaging machines, with the body designed by Alberto Pratelli, some of them with the supervision of one of the authors (Mario Salmon).

1.4 Fidelity to a Tradition

A periodic update of the product/service is an innovation strategy, called redesign. Redesign can have two totally different readings:

1. To revise the appearance. A redesign of this type, with the sole task of causing the user to feel new interest with non-functional aspects, is called restyling.
2. Vice-versa, redesign can not be a pure restyling. Extreme examples are when classic lines are proposed again on the market after some years, with substantial

(a) *The original design of a chair of the thirties'*

(b) *The redesign of the chair of 1956 by Eero Saarinen (melting metal covered with plastic)*

Fig. 1.13 An example of redesign of a chair. Modified from [19]

(a) *The original design of a FIAT 500 of the fifties (1957-1975). Chef-Designer D. Giacosa*

(b) *The a totally new redesign of a Fiat 500 in year 2007*

Fig. 1.14 An example of redesign of a small car

improvements of function. This happens often to vehicles, in which, however, the transformation can be remarkable in enlarging the market share, maintaining fidelity to an old design concept. Design can achieve dramatic improvements in critical areas (such as cost, quality, service, and response time), e.g. through an in-depth use of information technology and modern technologies [37].

Example 1.5 (*Redesign for Innovation*)
Some examples may clarify the meaning of the concept of redesign, in the case of limited production (of furniture) and in the case of mass-produced industrial objects, Figs. 1.13 and 1.14. An example of fidelity of a design is the case of the *Leitz Camera*, which maintains its visual appearance after more than 90 years, with total technical renovation, maintaining some characteristics of initial usability, Fig. 1.15. An other example is the *Vespa* motor-scooter design [34], Fig. 1.16. This case is much more

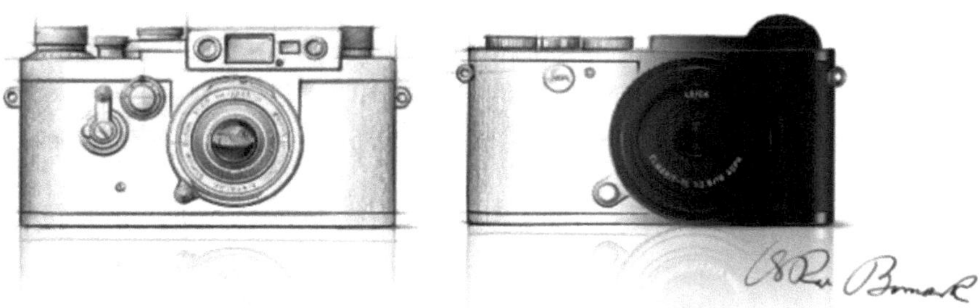

Fig. 1.15 An example of a re-design of an old camera: from Leica CL Catalog: an Oskar Barnack's Legacy (1925–2018). The old camera on the right, the new camera on the left

(a) *Original design of a Vespa (1946-) in the sixties of the last century. Chef-Engineer Designer: C. D'Ascanio*

(b) *The redesign of the Vespa in recent years*

Fig. 1.16 Another example of design of the Vespa motor-scooter

than a redesign. Formal values are kept as distinctive of this product, after more than 70 years and also the type of structure that obeys a principle of structural form, (the shell structure does not need other frames), even if all the inner mechanical, electrical and electronic parts are totally renewed.

1.5 What We Have Learned from This Chapter

Up to this point, we have introduced and utilized some basic concepts:

- **Products** that are not only physical objects, consumables, industrial artifacts, but also software or immaterial goods (know-how, patents, etc.), utilities (public services, theater, health ...) and materials from continuous processes (steel, oil, ...). Then in what we have learned, the word *product* can be coupled or substituted with the word *service*.
- **System** that is a set of connected things or parts forming a complex whole, in particular a set of things working together as parts of a mechanism or an interconnecting network or, in general, an organized scheme or methods.

After these definitions, we examined the design concepts from different points of views:

- Theoretical nature of Design.
- Definitions of Design concepts and activities done by the Design Communities.
- Real examples of products with good or not good design interactions.

To distinguish the tasks of Engineering Design from those of Industrial Design is not always straightforward and easy. Much is related to people skills and traditions.

To distinguish these two professional activities that have their meeting point on the product/service, i.e. the result of design, we followed this method: First we analyzed the differences in practice behavior with an observational study and a theoretical difference was identified. Then we analyzed their working process, based on a theory in the literature. At the end of this presentation, it is worth quoting a judgment at the ICED conference in 2009, reported in a review paper [31], that criticizes a general trend in design research and, vice-versa, points out some aspects that should deserve more attention:

> In that conference, little was said about service and product-service systems, little research was reported that bears on the detail design phase and even less that concerns the very important area of development. The study showed that while many papers reported research that models some phenomena or presents new ideas for methods, there was relatively little empirical back-up through observation of design in practice or through survey.

We conclude that a good interaction is recognizable in products intended for final users as well as in products intended for industrial customers.

References

1. ABET (1998) Accreditation board for engineering and technology. http://www.abet.org/
2. ABET (2000) Definition of engineering design by ABET. http://www.me.unlv.edu/Undergraduate/coursenotes/meg497/ABETdefinition.htm
3. ABET (2012) Criteria for accrediting engineering programs. http://www.abet.org/uploadedFiles/Accreditation/Accreditation_Step_by_Step/Accreditation_Documents/Current/2013-2014/eac-criteria-2013-2014.pdf. Accessed February 12, 2013
4. ADI (2016) Associazione per il disegno industriale. http://www.adi-design.org/homepage.html
5. AIGA (2016) The professional association for design. http://www.aiga.org/design/
6. Al-Muhammad M (2010) Guidelines to coop student and technical report writing. http://faculty.kfupm.edu.sa/COE/marwan/richfiles/Engineering_Design_Guidelines.pdf
7. BEDA (2016) Bureau of European design associations. http://www.beda.org/
8. Bokardo (2016) Principles of user interface design. http://bokardo.com/principles-of-user-interface-design/
9. Bonsiepe G (1983) Teoria e pratica del disegno industriale, 3rd edn. Feltrinelli Milano, new edition 1993
10. Bonsiepe G (1994) Del Objeto a la Interfase: Mutaciones del Diseño. Ediciones Infinito Buenos Aires. ISBN: 9789879637067
11. Bonsiepe G (2000) Design as a cognitive tool: the role of design in socialization of knowledge. In: Politecnico (ed) Design plus research, Politecnico di Milano, Ed. Politecnico, Piazza Leonardo da Vinci, Milano
12. DGTF (2016) German society for design theory and research. http://www.dgtf.de/english
13. DN CS (2016) International institution for production engineering research. http://www.cirp.net
14. DRS (2016) Design research society. http://www.designresearchsociety.org
15. Dym CL (1994) Engineering design : a synthesis of views. Cambridge University Press, Cambridge
16. Falzoni Gallerani V (2013) Paolo Stanzani: genio e regolatezza, vol 1, 1st edn. ASI Service srl Torino
17. Faulkner P, Runde J (2009) On the identity of technological objects and user innovations in function. Acad Manag Rev 34(3):442–462
18. de Figueiredo AD, (10–12 Nov 2008) Toward an epistemology of engineering. In: 2008 workshop on philosophy and engineering. The royal academy of, engineering London. http://ssrn.com/abstract=1314224
19. Freddi A (2004) Imparare a progettare, vol 1, 1st edn. Pitagora Editrice Bologna
20. IASDR (2016) International association of societies of design research. http://www.iasdr.org/node/1
21. ICED (2017) International conference on engineering design. https://www.designsociety.org/
22. ICSID (2013) History. http://www.icsid.org/about/about/articles33.htm
23. ICSID (2015) Industrial design last definition. http://www.icsid.org/about/about/articles31.htm
24. IDETCIE A (2016) International design engineering technical conferences. https://www.asme.org/events/idetccie
25. IDSA (2009) Definition of industrial design. http://www.idsa.org/education/what-is-industrial-design
26. Jemenez G (2014) Writing as social practice in engineering. In: Views from a cross-disciplinary study, John Benjamins
27. Kim KM, Lee KP (2010) Two types of design approaches regarding industrial design and engineering design in product design. In: Proceeding of international design conference, Design 2010, pp 1795–1806. http://dpl.kaist.ac.kr
28. Lidwell W, Holden K, Butler J (2003) Universal principles of design. Rockport Publication, 1
29. Lindbeck JR (1995) Product design and manufacture. Prentice-Hall Inc, Englewood Cliffs New Jersey

30. Liu S, Boyle IM (2008) engineering design: perspectives, challenges and recent advances. J Eng Des 20(1):7–19
31. McMahon CA (2012) Reflections on diversity in design research. J Eng Des 23(8):563–576
32. Osterwalder A (2010) Users vs customers. http://businessmodelalchemist.com/blog/2010/07/users-vs-customers.html
33. Pahl G, Beitz W, Feldhusen J, Grote KH (2007) Engineering design: a systematic approach, 3rd edn. Springer, London
34. Piaggio (2016) Homepage piaggio. www.it.piaggio.com/
35. Pratelli A, Salmon M (1981) Un esempio di integrazione di design, meccanica, elettronica nella progettazione di una obliteratrice. Il Progettista industriale (5)
36. Quarante D (1992) Diseño Industrial, vol 1 and 2. Enciclopedia del Diseño CEAC S.A. Barcelona
37. Reengineering in Business-Dictionary (2016) Business process reengineering. http://www.businessdictionary.com/definition/business-process-reengineering-BPR.html#ixzz46ZwbxniK
38. Sacchetti V (2009) Il Design in tasca, vol Ottagono Small, 1st edn. Cooperativa Compositori, Bologna and Feltrinelli
39. TRL (2012) Definition of technology readiness levels . https://esto.nasa.gov/files/trl_definitions.pdf
40. TRL (2015) Technology readiness levels. https://www.nasa.gov/directorates/heo/scan/engineering/technology/txt_accordion1.html
41. Usability-gov (2016) User interface design basics. https://www.usability.gov/what-and-why/user-interface-design.html
42. VDID (2016) Verband Deutscher Industrie Designer e.V. URL http://www.vdid.de/, this organization is a sister organization to the industrial designers society of America (IDSA) through the international council of societies of industrial design (ICSID)
43. Vitruvius (2006) The ten books on architecture by Vitruvius Pollio. http://www.gutenberg.org/ebooks/20239?msg=welcome_stranger
44. Vitruvius Pollio transl Morris Hicky Morgan (1960) The ten books on architecture. Courier Dover Publications. ISBN: 0-486-20645-9
45. Walke M (1989) Ducati Desmo: the making of a masterpiece. Osprey Publishing. ISBN: 0-85045-879-X

Chapter 2
Design Methods

Abstract It is customary to refer the expression cognitive science, since the end of the seventies, to all the disciplines that relate to the interdisciplinary scientific study of the mind and its processes. Cognitive science focuses on how information is represented, processed, and transformed (in faculties such as perception, language, *memory*, attention, *reasoning*, and *emotion*) in human and animal minds and machines. Let's apply these concepts to the analysis of the methods that are developed in the design process. The cognitive theory emphasizes the relationship between design thinking and memory. The tools for memory management are writing and drawing actions necessary to transfer experiences and mental processes to new subjects involved in the design process. Basically, we can say that design is a technical but also a humanistic process because it is related to a human problem solving process. Design in its most effective form is a process, an action, more than an object. It is a protocol for solving problems and discovering new opportunities. Every design process has common elements. Over the last fifty years, methods for understanding the nature of design have been discussed and proposed. While the design processes are distinctive in specific ways depending on the different fields (mechanical, precision engineering, electronic and software engineering, etc.), a common cognitive base of different methods exists.

2.1 The Cognitive Nature of Design

A subtle issue of *cognitive theory* must be pointed out: a method is always a *consequence* of an experience and not an *aprioristic* position[1] when dealing with a new problem. The reason is that any method derives from a reflection of the memory on human experience [21]: it starts with a perception of a problem and ends with some kind of related solution.

[1] Aprioristic = a priori reasoning, arguments, or principles.

A. Freddi and M. Salmon, *Design Principles and Methodologies*, Springer Tracts in Mechanical Engineering, https://doi.org/10.1007/978-3-319-95342-7_2

Table 2.1 Four macro-phases developed into eight phases

Requirements	System design	Detail design	Production
1. Product planning	3. Conceptual design	5. Embodiment design (prototype development)	7. Production start-up
2. Task clarification (specifications development)	4. First prototype	6. Detailed design (prototype verification)	8. Market follow-up

When the optimal solution of a problem is found, if retrospectively analyzed, an unavoidable need to move towards this solution becomes evident; only at the end of the process, can this path become a method, (i.e., literally, *meta-odos = the road through which*), in a way that takes the shape of inevitability. After this route has been completed, it seems, in fact, that the path by trial and error is the only right one [6, 12].

Every method tries to transform the complexity and variety of the design process into a systematic (which does not mean algorithmic) procedure. The most effective methods follow a descriptive principle They are suited to complex processes of the human mind [9, 23, 32, 33].[2]

Every method proceeds from the general to the specific, from *Requirements* to *Detail Design*. After this, the phases of manufacturing and external vendors, use and recycling take place. These phases are not discussed in this book.

We can deploy the three essential steps of a product development process: analysis synthesis evaluation in a different number of steps. E.g. in four steps:

1. Requirements
2. System Design
3. Detail Design
4. Manufacturing

or, in a more detailed analysis, in eight logical phases: Table 2.1, where Market Follow-up and Recycling are also considered phases of a product development process.

[2] The meaning of a descriptive approach is clarified by linguistic: "1. Descriptive grammar: the systematic study and description of a language. Descriptive grammar refers to the structure of a language as it is actually used by speakers and writers. 2. Prescriptive grammar: a set of rules and examples dealing with the syntax and word structures of a language, usually intended as an aid to the learning of that language. Prescriptive grammar refers to the structure of a language as certain people think it should be used" [18].

2.1.1 Common Origin: The Cartesian Method

René Descartes sought to ascertain the true method by which to arrive at knowledge, with special reference to what he considered exact sciences. This is a suitable starting point to analyze the development of the ideas that have led to modern design methods, both for engineering and for industrial design. The basic concepts of the Cartesian method are in a famous text, that is summarized here in its most relevant concepts [4, 7].

> [....], so I thought that in place of the large number of rules that make up logic I would find the following four to be sufficient, provided that I made and kept to a strong resolution always to obey them.
>
> 1. The first was never to accept anything as true if I didn't have evident knowledge of its truth: that is, carefully **to avoid jumping to conclusions** and preserving old opinions, and to include in my judgments only what presented itself to my mind so **clear and distinct** that I had no basis for calling it in question.
> 2. The second was **to divide each of the difficulties** I examined into **as many parts as possible** and as might be required in order to resolve them better.
> 3. The third was to direct my thoughts in an orderly manner, by starting with the simplest and most easily known objects in order to move up gradually to knowledge of the most complex, and **by stipulating some order even among objects that have no natural order of precedence.**
> 4. And the last was to make all my **enumerations** so complete, and **my reviews** so comprehensive, that I could be sure that I hadn't overlooked anything.

These concepts are the basis of every design method. Cognitive theory offers a natural extension of the method, since it highlights the role of imagination and experience, that was not the first concern of Descartes.

2.1.2 An Evolutionary Thinking: One Thing Leads to Another

A good synthesis between the rigorous fundamentals of the Cartesian method and the results of cognitive sciences is *Bruno Munari's method*.[3]

The rationale of this method is the following:

- Given a problem, the designer must proceed from the general to the specific, from outline proposals to detail design, avoiding seeking a solution by finding an idea (I) in the context of pure imagination.

---

[3]One of the last members of the Futurist generation, 1907–1998, Bruno Munari was among the most inspirational designers of all time [22].

- The solution of a design problem is never result of an impromptu invention, but rather of a cognitive process that supports a specific method, for keeping control of the context in which solutions operate and the technological environment in which the innovative idea should take form and be realized.
- Design is not finding a solution in an abstract way, but rather is the outcome of a procedure, where experience, (i.e. memory), rationality, (i.e. science results) and a contribution of imagination (i.e. invention) interact.
- The analysis of existing solutions is essential for positioning and judging new ideas, eliminating the drawbacks of the known solutions and improving the existing ones. This point may be a cause of dispute: how can you innovate starting from known solutions? In fact, the human being always moves from known solutions to new ones and an attempt to realize totally abstract ideas can not bear fruit.
  The present primer book differs from others based on a total abstraction: the book is a first reading and tries to facilitate a basic approach to design.

Figures 2.1 and 2.2, in their original form in the Italian language, summarize the sequence of a series of steps, that must be read in his way:

- Pose a problem (P) and search for a solution (S), **avoiding jumping to conclusions,**[4] and going directly to a solution.
- If we look at the evolution of thought during the design phase, we can observe that any solution is a choice between several possibilities that, in turn, leads to sub-functions (or lower-level functions) of which there are also more than one. This corresponds **to dividing each of the examined difficulties** into **as many parts as possible**, as Descartes said. The choice is one of several possible ones and so on.
- It is necessary to define the type of solution: temporary or permanent, sophisticated or simple, low innovation or very innovative solutions.[5]
- Don't search for an idea, but first define the problem (DP), clarifying the real customer needs: so the search for a *abstract idea* (I) is (indefinitely) postponed.

2.1.2.1 A Case Study by Munari's Method: Design of an Interior Lamp

This is the case study of a lamp design for a domestic use.

Whatever the problem, it can be disassembled into its components CP (= components of the problem) :(ABC), (DEFG), (HILM), then in *sub-components* (A, B, C, …, M). Figure 2.3 shows the summary of the method, with the steps sequence. The design phases are the following, Fig. 2.2:

[4] Compare this sentence with the previous Cartesian words.

[5] This observation is not present in the exact sciences.

Problem steps	2nd	3rd	4th	5th	6th	7th	8th	9th	10th	11th
P= Problem	P	P	P	P	P	P	P	P	P	P
S= Solution	DP	DP	DP	DP	DP	DP	DP	DP	DP	DP
	I	Compone nts	CP	CP	CP	CP	CP	CP	CP	CP
	S	I	Data Collection	DC	DC	DC	DC	DC	DC	DC
		S	I	Data Analysis	DA	DA	DA	DA	DA	DA
			S	I	Creativity	C	C	C	C	C
				S	S	Tecnology Materials	MT	MT	MT	MT
						S	Experimen tation	SP	SP	SP
							S	Model	M	M
								S	Verification	V
									S	Techn. drawings
										S

Fig. 2.1 Steps of Munari's method

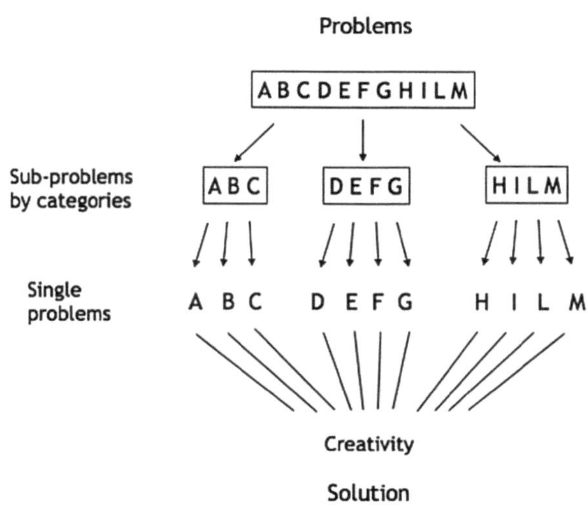

Fig. 2.2 The 4th step of Munari's method. P = problem, S = solution, DP = *Definition of the Problem* (Clarification), I = Idea, CP = *components (and sub-components)* of the problem. On the right side: dis-assembly phase of the main problem (analysis and synthesis phases). Modified from [22])

- The *reductionism* principle is applied to **disassemble a problem into its component parts**.

  In our example the sub-problems must be presented as **clear and distinct from one another**.
- This result can be reached utilizing analysis tools such as those formalized in the following questions and seen again in the next chapters:

1. Who uses it, i.e. how to use it.
2. Where it is located.

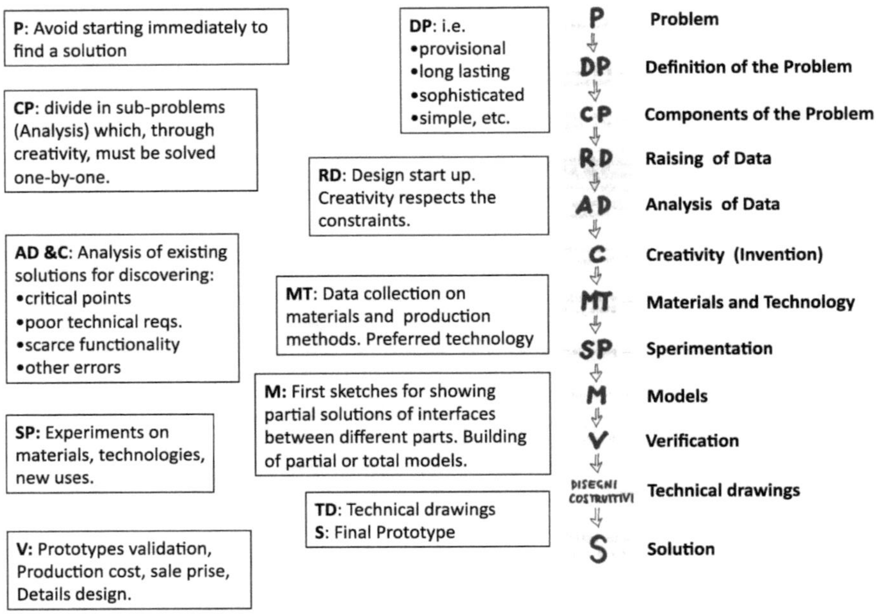

Fig. 2.3 Synthesis of *Munari's* method for the solution of a design problem (original drawing with literal translation of the symbols)

3. When it is used.
4. Why it is used.
5. What kind of light.
6. What kind of adjustment.
7. What kind of material.
8. Who knows the proper technology.
9. Where it is stored.
10. What prefabricated parts (or not) are on the market.
11. What the shape is.
12. How it is transported and packaged.
13. How much it must cost.

The subsequent steps of the method are the following:

- Get all the data (RD = data collection) necessary to study these CP (components of the problem or *sub-problem* or, for complex design, *sub-sub-problems*), one by one, **by stipulating some order even among objects that have no natural order of precedence**. (This is another Cartesian observation).
- Before focusing on creativity for solving the singular component of the problem (A, B, C, …), collection of catalogues, data from handbooks for similar designs and a list of *off-the-shelf* components are recommended.

The data we must collect are catalogs of similar lamps, types of light bulbs on the market, names of external suppliers etc.
So the search for an *abstract idea* (I) is not necessary yet, and must be postponed once again.

- Analyze data on solutions adopted by competitors and in similar products offered on the market. The role of *creativity* is limited and controlled by information received from the data analysis. It is a sort of *driven mental activity, not a free artistic idea.*
We analyze the various types of similar products on the market to try to discover defects (lack of aeration, dispersion and absorption of light, etc.). Generally, suggestions provide clarification about what to do and what not to do.
Figure 2.1 shows the second part of the method: The first step on the left side is the 5th-step of the overall method: "raccolta dati" means raising of data (*data collection*). The following step (the 6th), develops the *analysis of data.*
The 7th step introduces a controlled role of *creativity*, the 8th *materials and technology*, the 9th the *experimentation*, the 10th *mode construction* and the 11th the *verification.*
Creativity, i.e. inventions must be limited to the phases where they are effectively necessary.
In this way, the search for a revolutionary idea-solution is not necessary and is indefinitely postponed in the design process, until the final solution naturally emerges.[6]
- Choose materials and technologies. The choice of technology must be adapted and consistent with the manufacturer know-how.
- Test materials and new technologies. Experiments on materials and technologies can give information on new uses of a product.
- Draw and build models and define the first prototype. The hardest designer role is to reconcile the various solutions with an overall design. *The target is to concentrate several functions on just a few components.*
- Verify prototypes.
- Develop the embodiment design and all the detail drawings. (We will see examples of the last points in next chapters.)

2.1.3 Counter-Example

Criticism for all design methodologies [10] derives from observation of the apparent holistic nature of design activity. This could be true for the design of elementary products in which a non-methodical path of the mind (as a pure aesthetic experience) can prevail. With a criterion of pure formal freedom, simple conceptual projects on lamps were carried out, following free associations of ideas based on intuition.

---

[6]This is in fact the main message of the method: do not search for new abstract ideas out of context!.

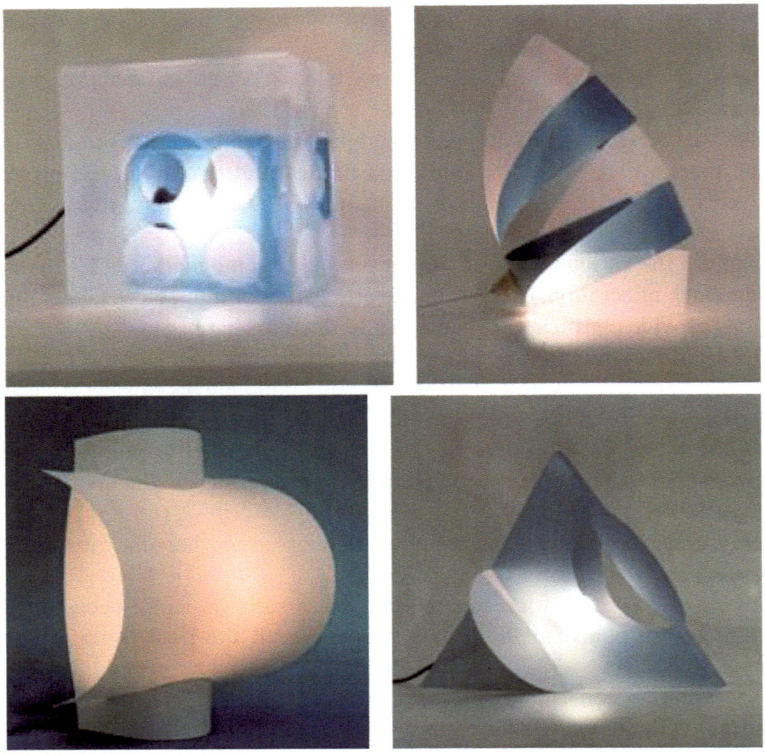

Fig. 2.4 Simple solutions for lamps, obtained by plastic colored foils. Works of students (L. Damerini, M. D'Angelantonio, M. Fedriga in years 1999–2000 (Tutor: V. Sacchetti). Modified from [11]

Example 2.1 (*Other Designs of an Interior Lamp*) A student team was asked to suggest solutions for table lamps, only by providing colored plastic sheets and bulbs [25]. The results are shown in Fig. 2.4. We can frankly admit that, in some cases, the anomaly of certain solutions and the role of unpredictable inventiveness can suggest the best solution totally outside a systematic approach [29, 35].

2.2 VDI Guidelines

A methodical point of view is represented by the philosophy of VDI[7] guidelines that guide in developing a design process. The VDI-GPP Society (GPP = Product and Process Design) with its technical divisions offers knowledge regarding the design

[7] Verein Deutscher Ingenieure.

of products and processes and their optimization in terms of quality, time and cost-benefit ratio for all industries. The final task of the guidelines is to transform scientific results into real design practice.

The VDI Society consists of eight technical divisions:

- Information Technology.
- Sales Engineering and Product Management.
- Value Management/Value Analysis.
- Project and Process Management.
- Reliability.
- Vibration Technology.
- Product Development and Mechatronics.
- Mechanism Design and Machine Elements.

A catalog of all the valid VDI Guidelines and specifically the VDI-GPP manuals appears every six months. They are produced in independent committees of leading experts from industry, science and administration. They form a working aid with clear evaluation and assessment criteria for practical daily work. The VDI-Guidelines 2221 deal with general (industry-independent) methodological development and construction of technical systems and products. The Guidelines started in 1954 as a methodological aid to inventiveness and to the designer's task, highlighting common aspects of invention processes. Their task was to encourage the designer to liberate him/herself from any prejudices and to pursue the paths of solution search that are better than the existing ones, weighing up technological experiences of the past but, at the same time, considering new ideas and thought associations, even if they seem absurd at the first instance.

Subsequently, the method evolved in the direction of the analysis process rather than analysis of the inventive concept. In 1993 the guidelines adopted their modern form, with the aim of a general method suitable to design technical systems and products, with an abstract approach, independently of the branch. Pahl & Beitz's book largely reflected the spirit of this approach (one of those authors has been one of the main authors of these guidelines since 1973), and develops an authoritative approach to systematic design. Systematic design methods consider design as a result of a reductionist logic, i.e. an articulation of abstract functional relations with which principles of concrete solution must be associated. In this way, almost theoretically, a complex problem can be solved without conceptual dependencies on past (already applied) solutions. Special emphasis is placed on the iterative nature of the approach and the sequence of the steps must not be considered rigid. Some steps might be omitted, and others repeated frequently. Such flexibility is in accordance with practical design experience and is very important for the application of all design methods. The guidelines introduce several important principles and methods that can be regarded as strategies that help the development of design processes, even of different sub-sectors of the industry and of most other companies in the economy. A systematic design approach has become only partially accepted, just thanks to what has been done in terms of international standards, although with differences that reflect scientific and technical traditions of different industrial experiences [26].

It shows stages and results of any steps, i.e. the differences between thinking and acting phases. Figure 2.5 should therefore be regarded as a guideline to which detailed working procedures must be assigned.

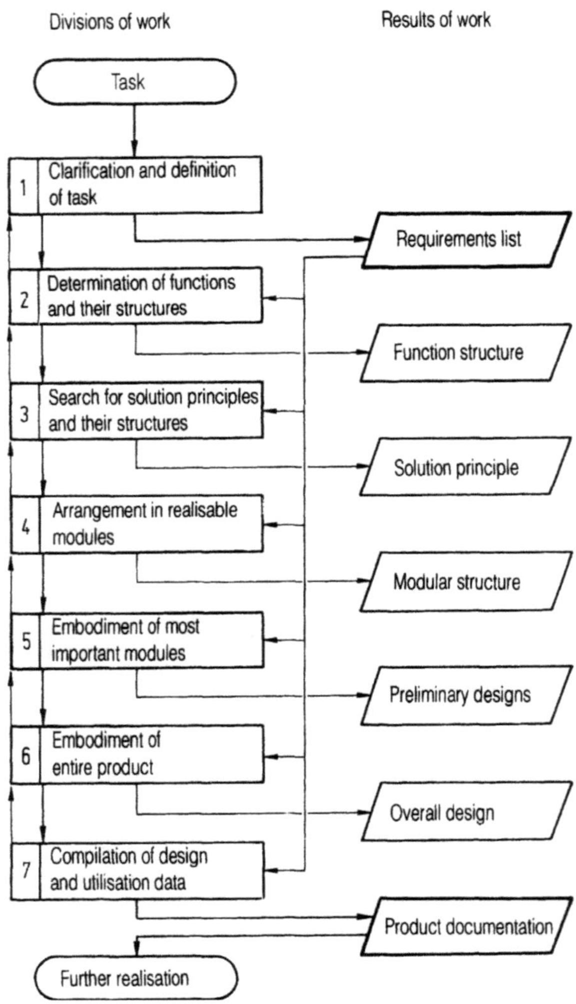

Fig. 2.5 General method for design from Dubbel [3, 16]

2.3 The Pahl and Beitz Method

This method [23] is a structured summary of systematic approaches. The method can be summarized in the following steps.

2.3.1 *Conceptual Design*

The specification as presented to the designer may not be comprehensive and often requires clarification and additional information. The users' need clarification (with all the relative tools) is a possible way to reach a complete specification to develop new products. In the event that conceptual solutions are already applied, this step can be skipped and we can proceed directly to the embodiment and detail design phases. This is the case of many industrial machines because they have a clearly identified customer who clarifies directly to the designer his goals and his desires, which, in the majority of cases, translate into improvements in the efficiency and productivity of the machinery.

2.3.1.1 Searching for the Abstract Functional Structure

If we think about a technical function, we must admit that it may be realized by very different *solutions*. While the function expresses a general and abstract relationship of a system, a subsystem, or even of a single component, the way to achieve it can be extremely varied. See Fig. 2.6.

The simple example shows how the same abstract function "to cut" can be realized by two totally different technical solutions.

We observe that the same overall function is realized with a different basic technological process (shear stress or compression stress), the force amplification (by means of levers) is different, and the return to reach a new starting point is allowed respectively by rings for an inverse force application or by a spring action. The nail clipper also offers a second finishing function (with a file), a third function of compacting the tool with the lever rotation, and a fourth function fulfilled by a hole for linking it to other parts. The same Figure shows the functional structure written in two different ways:

- On the left the functions are written in a form that is valid for both systems while,
- On the right the functions are written in extreme generalization, with only four actions (change, transport, vary and connect), that, according to the Pahl & Beitz analysis, are valid for any system.

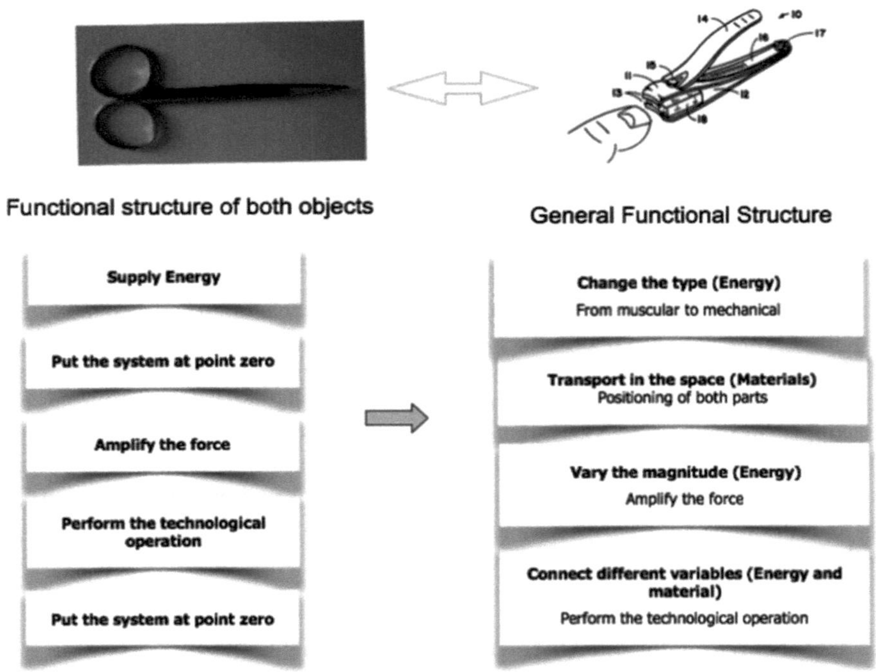

Fig. 2.6 There is not only one way of doing things

The systematic approach needs implementation of the following actions:

- Identification of the overall functional structure.
- Reduction of overall function to linguistic relationships between three basic entities: *energy, materials and signals*, (signals are the carriers of information). The basic relationships are expressed by means of a very limited number of imperative verbs:

1. Change the type.
2. Vary the magnitude.
3. Connect different variables.
4. Carry in space.
5. Store in time.

- The overall function can be sub-divided at increasing levels of division, (first, second, third, etc.). It involves breaking down the overall function into sub-functions until the sub-function becomes clear and simple in Cartesian meaning.

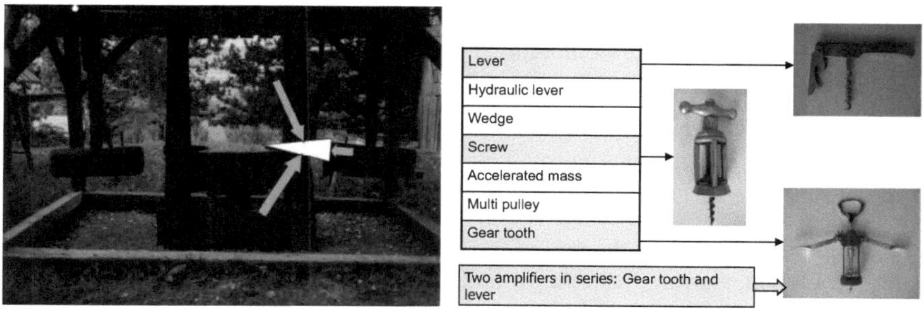

Fig. 2.7 Example of a rudimentary farming device for pressing seeds with double amplification system (on left) and examples of solution principles as Force Amplifier (on right). Modified from [11]

2.3.1.2 Searching for Solution Principles

A useful exercise is to get used to "interpreting" the devices and the technical systems, identifying their functional structure immediately, and the adopted solution principles, see examples in Fig. 2.7. Firstly, look at this old press for obtaining seed oil in a farmer environment, Fig. 2.7 on the left side. If we interpret the device through functional analysis, we find that the rudimentary press consists of two systems of forces acting in parallel, each formed by two amplification systems acting in series, the pendulum, (energy generator but also force amplifier due to the acceleration of the hammer mass), and the wedge, that is a second force amplifier.

Similarly, we can recognize a force amplifier in a common device based on the principle of screw, lever or gear tooth.

2.3.1.3 Combine the Solution Principles into Concept Variants

A combination of solution principles can be regarded as a sort of a *conceptual functional variant*. An effective, visualization tool for comparing different combinations is given by the so-called *morphological matrix*. After a stage of pure freedom in combining different solution principles of every function (without control of their consistency), a preliminary selection is conducted in order to eliminate absurd or intrinsically not safe combinations.

2.3.1.4 Evaluation of Concept Variants Using Technical and Economic Criteria

This is the beginning of a synthesis procedure, done through several test screenings. A process starts by evaluating and selecting the most promising combinations, faced with technical and economic criteria.

The development of preliminary layouts and form design, after further refinements, leads to a preliminary embodiment design.

2.3.2 *Embodiment Design*

The embodiment design is the part of the design in which the product plan is specifically developed, starting from the variants selected by the conceptual design.

The embodiment design, like the conceptual design, is developed according to the following logic:

- Cataloging of different solutions.
- Selection of the best solutions.
- Elimination of errors.
- Optimization of the best solutions.

The same tools introduced for conceptual design can be used. This stage is divided into a preliminary design and a detail design phase.

2.3.2.1 Develop a Definitive Layout and Check if All Requirements Are Met

The layout of each selected concept variant is verified from the point of view of function, strength and spatial compatibility and the definitive variant is chosen, checking that all requirements are met.

2.3.3 *Detail Design*

The exploration of options does not finish with a conceptual solution but extends to the physical realization of the products. Again, the principle of systematic design should be applied. The principles of *clarity and simplicity* must be applied, respecting the standards on safety that have been developed according to, e.g. the Machinery Directive Harmonized Standards or similar.

2.3.3.1 Documentation

Detail design also includes the production of final design drawings which can be used to manufacture the first prototype and subsequently arranged for mass production.

2.4 Other Systematic Approaches

Other methods have been proposed in the extensive literature on this topic, for the systematic development of the various design phases. A wide range of references and a general discussion and comparison is given by [37]. Other relevant approaches and schemes can be found in [8, 33]. What is common in all the methods is that a market analysis phase (or a clarification phase) should precede any specification, particularly when the design is developed for a mass product, intended for a large market where final users' needs and attitudes must be identified. Limiting the analysis of the design concept to a few essential categories, a design process is developed in four phases [9]:

1. Task Formulation.
2. Functional Phase.
3. Form Design.
4. Forming Results.

There are no major substantial differences with other systematic approaches: e.g., referring to one of the most known methods (Pahl & Beitz), the Functional Phase corresponds with the Conceptual Design phase and the Form Design Phase is similar to the Embodiment Design phase.

2.5 Systematic Engineering Design in an Industrial Context

As we noted, the Pahl & Beitz method is generally geared to the logic of Problem-Solving. If we want to insert this method in the broader structure of the industrial product design, we should remember that the technical task is just the core of a larger system which includes different actors such as project management, company strategies and market analysis. A company's structure is organized in several areas [13] where Engineering is only a part. The problem-solving logic develops into a product-oriented logic in the production process.

We can summarize this concept with Fig. 2.8a, in which the engineering analysis in Fig. 2.8b is the main piece to insert in the general product development context.

2.6 Design Standards

Every discourse on a method must take into account standards and guidelines. Each production process, also the craft of a single craftsman, requires the planning of actions according to a protocol specification. While in the handicraft case, this route can be in the mind of just a few persons, or written in a simple document, in the case of an industrial product, this path is more complex, with decision-making, control and review steps, and needs to be completed without eliminating any steps.

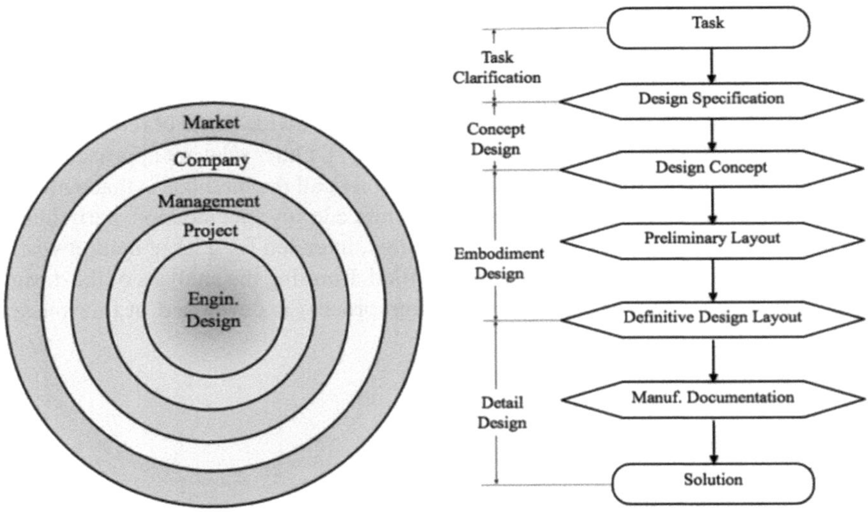

(**a**) *General Scheme.* (**b**) *Pahl & Beitz Process Flowchart .*

Fig. 2.8 Engineering design context and the Pahl & Beitz process-flowchart [13], (from a Springer Book)

It is also important that the whole system devoted to the development of a product always works according to constant criteria and clearly defined specifications. For this reason, a Quality Management System (QMS) must be introduced as a necessary reference in any production system.[8]

The QMS ISO-9001 standards are the result of international expert consensus and, which is important to understand their philosophy, are not applicable to the variability of products, but to the company that produces them. This is in fact a necessary pre-condition for the quality of the products. The product quality is the task of the product design, since product quality has two main values and aspects:

- The characteristics of a product or service that have a bearing on its ability to satisfy stated or implied needs.
- A product/service free of deficiencies. In any case, conformity to International Standards is then important for potential consumers, because it helps reassure them on the safety and reliability of products and systems.

Both are definitions of the American Society for Quality (ASQ). This is the only standard that can be certified (although it is not a requirement) [2, 15].

[8]Even if Quality has some personal (not-quantitative) value since it is related to human experience, it must be defined, measured and applied.

2.6.1 *Quality Management System (QMS)*

> A quality management system (QMS), issued by the International Organization for Standardization, (ISO) is a formalized system that documents processes, procedures, and responsibilities for achieving quality policies and objectives. A QMS helps coordinate and direct an organizations activities to meet customer and regulatory requirements and improve its effectiveness and efficiency on a continuous basis [1]

ISO 9001:2015 is the standard for implementing this quality management system.[9] It specifies the requirements that each organization can use to develop its own quality programs. Consulting services or other equivalent consulting centers active in different Countries focus on advising organizations on how best to use the ISO Standard [1, 2, 17, 24, 27, 28, 30, 31, 34, 36]. The ISO 9001 does not indicate the way to implement the design process, but only *what needs must be included* in the process to ensure a proper design activity.

The main differences regarding the design process, contained in ISO 2001:2015 as compared to ISO 9001:2008, are the following:

- ISO 2001:2015 introduces the Annex SL structure, which outlines a common structure for all ISO management standards.
- ISO 2001:2015 has ten clauses, instead of the eight clauses of ISO 9001:2008. So the Operation clause becomes clause 8, instead of 7.
- The term *product* is replaced by *products and services*, so that standards becomes applicable for service-oriented organizations too.

2.6.2 *Operation Planning*

The operational plan is the task of cross-functional leaders to select strategies and customer requirements. The first step of this process is to identify the tools you will use to accomplish this task. One of these is based on asking questions of this type:

- What: addresses the strategies to be chosen.
- Who: personnel charged with accomplishing what.
- When: a time schedule for each activity.
- Needs: resources and controls to perform each activity.

The task of ISO 9001:2008-Clause 7.3, now ISO 2001:2015-Clause 8.3: *Design and Development of Products and Services* is to achieve a clearly defined design process for improving organization, collaboration and communication between all stakeholders.

---

[9]ISO standards are reviewed every five years by the standards committee responsible for their relevance, suitability and effectiveness. A systematic review of ISO 9001:2008 was done according to the general criteria of ISO directives.

2.6.2.1 Design and Development for Products and Services

Design is defined as a collective term of all the activities referring to the choice of the look and feel of the product and service. Design is not a result of an idea but of a process organized in several steps to be completed in logical sequence [22].

Development is the process of analyzing return of investment, marketing opportunities, identifying inputs and outputs and creating and modifying the product and service [17]. Development is done by a project team. Various activities such as reviews, verification and validation, involvement of customers, and interactions between company functions complete the process.

Design and development are the key to success, since they impact directly on the selection of the customer needs, the organization structures and the reviews and verification of the products. The design and development phase must be planned according to the following steps[10] (Fig. 2.9):

1. Design and development planning.
2. Design and development inputs.
3. Design and development outputs.
4. Design and development review.
5. Design and development verification.
6. Design and development validation.

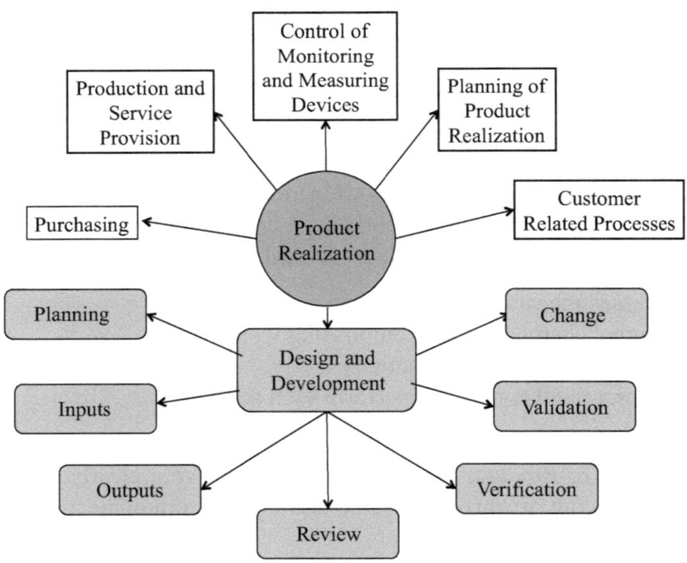

Fig. 2.9 Scheme of a design process according to ISO 9001

[10] Clause 7.3 (ISO 9001:2008) or Clause 8.3 (ISO 9001:2015) that integrates the previous one. In the 2015 edition, products and services are totally assimilated.

7. Control of design and development changes.

The clauses are shown in Fig. 2.8.

Planning

The organization shall plan and control the design and development of products. The planning shall include the following steps:

- Stage reviews.
- Verification and validation activities.
- Responsibilities and authorities for design and development.
- Interfaces between different groups involved in design and development to ensure effective communication.
- Output, as the design and development progresses.

Inputs

Inputs relating to product requirements shall be determined and shall include:

- Functional and performance requirements.
- Statutory and regulatory requirements.
- Information derived from previous similar designs.
- Other essential requirements for design and development such as impacts and implications of failures.
- Resolution of conflicting inputs.

Outputs

The outputs shall be in a form suitable for verification against the design and development input and shall be approved prior to release. The outputs shall include the following points:

- To meet the input requirements for design and development.
- To provide appropriate information for purchasing, production and service provision.

Note: Information for production and service provision can include details for the preservation of product.

Review

ISO 9001-Clause 7.3.4 Design and development review of ISO 9001:2008 shall complete systematic reviews:

- To evaluate the ability of the results of design and development to meet requirements.
- To identify any problems and propose necessary actions.
- To propose follow-up actions.

Participants in such reviews shall include representatives of functions concerned with the design and development stage(s) being reviewed. Records of the results of the reviews and any necessary actions shall be maintained.

Verification

For ISO 9001:2008 the organization shall verify that:

- Verification shall be performed in accordance with planned arrangements to ensure that the design and development outputs have met the design and development input requirements.
- Records of the results of the verification and any necessary actions shall be maintained.

Validation

For ISO 9001:2008 the organization shall control that:

- Validation shall be performed in accordance with planned arrangements to ensure that the resulting product is capable of meeting the requirements for the specified application or intended use, where known. Wherever practicable, validation shall be completed prior to the delivery or implementation of the product.
- Results of validation have been recorded.

Note: Verification and validation have distinct functions. Verification is a theoretical exercise designed to make sure that no requirements are missed in the design, whereas validation is a practical exercise that ensures that the product, as built, will function to meet the requirements. Together, they ensure that the product designed will satisfy the customer needs.

Changes

- Design and development changes shall be identified.
- Changes shall be reviewed, verified and validated as appropriate, and approved before implementation.
- The review of design and development changes shall include evaluation of the effect of the changes on constituent parts and product already delivered.

The application is followed by *Audit Checklists* that should cover the different areas of Clauses [14].

2.7 What We Have Learned from This Chapter

The Cartesian reflection is the first structured approach to a method and may also apply to design. Nevertheless, modern cognitive science offers a better knowledge of the mechanisms of the mind that are responsible for inventions and development of new ideas and make it possible to specialize the design method.

Standards and Guidelines are references for every design process. We considered two different types of Standards and Guides:

- ISO management system standards that provide a model to set up a quality management system. The **ISO 9001** does not offer ways on how to implement the design process, but only the necessary sequence of actions to accomplish a design task.
- VDI guidelines, such as the **VDI-Guidelines 2221** and other systematic methods which allow practical development.

Each design method relies on a set of principles that range from the most general, valid for all designs, to others, which relate to more specific types of design.[11]

We can try to distinguish the *principles of design* according to the following general scheme:

- *Universal principles* which refer mainly to the cognitive structure of the human mind [20]. They are general and must always be applied to industrial and to engineering designs. Issues relative to *How can I enhance the usability of a design?* and *How can I make better design decisions?* are examples of common and general interest.
- *Physical general principles* that belong mainly to the engineering conceptualization of the project and are related more to the functional and structural aspects. They are a guide with which to evaluate the *Solution principles*, consisting of physical as well as geometrical contents that satisfy the single sub-functions. These principles must be found through bibliographical and patent searches as well through analyses of natural and technical systems, analogies, measurements, and results from model tests. See examples in subsequent chapters.
- *Principles of Embodiment Design*. They are solution principles specifically developed for the type of embodiment design. So we have principles for mechanical design [3, 23], for reliability design [5], for industrial design [19, 20], etc.

This division of the principles in the three classes is not categorical. Often, in a conceptual stage, all the principles must be applied in their general meaning.

Any guideline benefits from specific industrial experience. The teaching of design methods requires abstract concepts to become common practice in the industrial environment. In order to do this, a task of adaptation and customization of the methods to different contexts is necessary.

The book does not deal with the final stage of design development and production (product industrialization) which are tasks of manufacturing phase. There are design principles suitable for this stage that are not illustrated. E.g.,

[11] All these types of principles have been progressively called up in the side of the pages, whenever they have been applied.

criteria for improving assembly (or dis-assembly), modular production with variety reduction criteria, when common parts can be designed and produced for families of products. These criteria are presented in specialized books.

The main steps of the systematic approach will be illustrated when design concerns especially adaptive or a variant solutions, through the case studies of the next chapters.

References

1. American Society for Quality-PO Box 3005 Milwaukee W (2017) What is a qms? iso 9001 & other quality management systems. http://asq.org/learn-about-quality/quality-management-system/
2. Barouch G, Kleinhans S (2015) Learning from criticisms of quality management. Int J Qual Serv Sci 7(2):201–216. http://search.proquest.com/docview/1686426979?accountid=9851, emerald Group Publishing Limited
3. Beitz W, Kuttner KH (1994) Handbook of mechanical engineering, vol 1, ebook edn. Springer-Verlag, London
4. Bennett J (2015) Discourse on the method of rightly conducting ones reason and seeking truth in the sciences. http://www.earlymoderntexts.com/assets/pdfs/descartes1637.pdf
5. Birolini A (2017) Reliability engineering, 8th edn. Springer
6. Bodei R (2011) Felici di fare conoscenza. http://www.ilsole24ore.com/art/cultura/2011-11-13
7. Descartes (1938) Discorso del metodo. Alberto Morando Editore
8. Dym CL (1994) Engineering design: a synthesis of views. Cambridge University Press, Cambridge
9. Dym CL (2012) Engineering design: representation and reasoning, 2nd edn. Cambridge University Press, Cambridge
10. Feyerabend P (1975) Contro il metodo: Abbozzo di una teoria anarchica della conoscenza. Feltrinelli, Milano
11. Freddi A (2004) Imparare a progettare, vol 1, 1st edn. Pitagora Editrice Bologna
12. French M (1994) Invention and evolution: design in nature and engineering, vol 1, 2nd edn. Cambridge University Press
13. Hales C, Gooch S (2004) Managing engineering design, 2nd edn. Springer
14. Hammar M (2015) ISO 9001:2015. http://advisera.com/9001academy/what-is-iso-9001/
15. ISO (2015) Quality management. http://www.iso.org/iso/home/standards/management-standards/iso_9000.htm
16. Jänsch J, Birkhofer H (2006) The development of the guideline VDI 2221—the change of direction. In: Proceeding of international design conference, DESIGN 2006, Dubrovnik-Croatia. Theory and research methods in design, pp 45–52. http://docentes.uto.edu.bo/mruizo/wp-content/uploads/VDI2221.pdf
17. Jarvis A, Palmes P (2016) ISO 9001:2015 : understand, implement, succeed! Boston: Prentice Hall, uL: Order in West Room (Not borrowable) (2016.9.1883)
18. Khedr M (2014) What's the difference between descriptive and prescriptive grammar? https://www.bayt.com/.../what-s-the-difference-between-descriptive-and
19. Kim KM, Lee KP (2010) Two types of design approaches regarding industrial design and engineering design in product design. In: Proceeding of international design conference, design 2010, pp 1795–1806. http://dpl.kaist.ac.kr
20. Lidwell W, Holden K, Butler J (2003) Universal principles of design. Rockport Publication, l

21. Marchis V (2012) Gli ingegneri delle rivoluzioni: industria o scienza?, vol 1, no 1, p 653. Clueb Bologna
22. Munari B (2006) Da cosa nasce cosa (One thing Leads to Another). Laterza Roma-Bari
23. Pahl G, Beitz W, Feldhusen J, Grote KH (2007) Engineering design: a systematic approach, 3rd edn. Springer, London
24. Rheinland T (2016) ISO 9001:2015 revision quality management of the future. http://www.tuv.com/en/corporate/business_customers/management_system/quality_management_systems/iso_9001_certification/iso-9001-2015-revision.html
25. Sacchetti V (2009) Il Design in tasca, vol Ottagono Small, 1st edn. Cooperativa Compositori, Bologna and Feltrinelli
26. Schmidt-Kretschmer M, Blessing L (2006) Strategic aspects of design methodologies: Understood or underrated? In: International design conference—Design 2006, Dubrovnik-Croatia. Theory and research methods in design, pp 125–130
27. Stein L (2012) Handbuch zur Einfhrung eines Qualitätsmanagement systems gemäss DIN EN ISO 9001: 2008. Diplomica Verlag, Hamburg
28. Systema-Consulting (2016) Ente per la certificazione della qualitá. http://www.scitalia.com/area-consulenza/certificazione-qualita-uni-en-iso-9001.html
29. Tae (2012) Theories about engineering and engineer theorists. http://theoriesaboutengineering.org/gerhard_pahl_and_wolfgang_beitz.html
30. Tricker R (2013) ISO 9001:2008 for small businesses, 5th edn. London: Routledge, uL: Order in West Room (Not borrowable) Available, (2014.8.2000)
31. Tricker R (2016) ISO 9001:2015 In Brief , 4th edn. Routledge
32. Ullman D (1992-2009) The mechanical design process, 4th edn. McGraw-Hill N.Y., ISBN-13: 9780071267960
33. Ulrich KT, Eppinger S (2008) Product design and development, 5th edn. Mc-Graw Hill, ISBN: 978-0073404776
34. Vology (2014) Gets design and development ISO 9001:2008 quality certification. Wireless News http://search.proquest.com/docview/1622167400?accountid=9851, copyright—Copyright 2014 Close-Up Media, Inc. All Rights Reserved
35. Wallace KM, Blessing LTM (2000) Observations on some German contributions to engineering design. In: Memory of professor wolfgang beitz. Research in Engineering Design
36. Wilson CL, John P (2016) Developing a knowledge management policy for ISO 9001: 2015. J Knowl Manag 20(4):829–844
37. Wynn D, Clarkson J (2005) Models of designing, vol I, pp 35–50. Springer

Chapter 3
Requirements and Specifications

Abstract Subjects involved in the design and production of goods have to build their decisions on market data. Collecting and organizing information is then the preliminary activity, necessary in understanding the complexity of the customer needs, often not "clear and distinct", but on the contrary, expressed in a confused and interrelated way. Users can play an active role that must be taken into account: user-driven design versus user-centered design is a controversial position that it is worth analyzing. The design process starts by the identification of the needs of potential customers and proceeds with conceptual design, then embodiment and finally detail design, and the knowledge and satisfaction of user real needs is crucial in any phase. The tools developed for acquiring this knowledge are far from being formalized (i.e. mathematically formulated) but, nevertheless, are consistent and based on principles of logic, cognitive sciences and on the human experience of all the subjects that collaborate in the design process. This great work of interpretation and translation leads, at the end of the conceptual design, to fundamental documents which are called "product (or service) specifications" and which are the starting points for all the following design phases.

3.1 Definitions

According to a general approach of this book, we can first collect and introduce some definitions found in the technical literature to highlight the common feeling of knowledge shared by the technical community on the product planning concept. A basic definition of product planning is the following:

> Product planning is *the process of creating a product idea and following through on it until the product is introduced to the market*, [25].
>
> Said another way, product planning is *the process of coming up with a business idea for an artifact, preparing the goods for production and then introducing them to the market*, [6].

This chapter aims to give an overview on product planning in the conceptual design phase.

A. Freddi and M. Salmon, *Design Principles and Methodologies*, Springer Tracts in Mechanical Engineering, https://doi.org/10.1007/978-3-319-95342-7_3

The necessity for product planning derives from the analysis of the life cycle of existing products in their life cycle (see next chapter). During any phase, a strong focus must be maintained on a number of key business areas such as development, financing, marketing, and manufacturing, so that the five areas of research for correct product planning are [18]:

- New market demands (functionality, fashion).
- New technologies.
- Possible technical superiority of competitors.
- Political and economic situation.
- Laws on recycling and environmental protection.

In the face of these constraints the main actions to be developed, are:

1. systematic search,
2. selection,
3. development,

of new product-ideas compatible and consistent with the company core competency.[1]

3.2 Systematic Search

The *voice of the customer* is the conventional term to describe stated and unstated customer needs (or express requirements). The market demands are captured in a variety of ways: direct discussion or interviews with potential customers, surveys, focus groups, customer specifications, observation, warranty data, field reports, etc.

Who really is the customer and who is the user of a company? An approximate difference is the following: customer refers to the party that pays for products and services, while end user is the party that uses the products and services. A lapidary definition: customers buy and users use. It could be that they are coincident but not necessary [23]. Without entering into this complex problem of the function of these subjects for the success of a product, are different types of customers: e.g., customers of products/services as final or intermediate users (external customers), but also members of the different business functions that are located downstream of the product development phases (such as manufacturers etc. (internal customers).

[1]Core competency is *a harmonized combination of multiple resources and skills that distinguish a firm in the marketplace* [22]. The “litmus test” for a core competency is that it is hard for competitors to copy or develop.

Table 3.1 Interview of a potential user of the product

Question	Description
Who?	Who uses the product?
What?	What is the use?
Where?	Where is the product used?
When?	When is the product used?
Why?	Why is the product used?
In which way, How?	How is the product used?

3.2.1 *Information Gathering*

The first action of carrying out a scientific research of the market is proper information gathering.

The interviews are the first tool in understanding the properties that the product must have to satisfy the needs and deliver its value. The respondents (interviewed) can be a small number of people or even individuals among current customers, competitor customers, and direct users of the same market segment.[2]

The simplest form of interview is done on the basis of *rhetoric*, according to the so-called *5W questions* method, Table 3.1. We have already met this type of interview in the Munari method introduction that was applied to the simple design of a lamp in an informal way.

The five questions plus how? question are fairly famous and an often-quoted way to think through problems.[3] The utility of the rule is to help synthesize the most important data or, said another way, the indispensable information.[4] Thus the 5W rule is used in problem solving questions and (with some modifications) in product planning [7].

The Five Ws: *When? Where? Who? What? Why?* plus a sixth *How?*[5] are questions whose answers are considered basic and useful for getting the complete story on a subject, for building scenarios and knowing contexts, Table 3.4 [20, 28]. The customer needs belong to four categories:

[2]A market segment is an identifiable group of individuals, families, businesses, or organizations, sharing one or more characteristics or needs in an otherwise homogeneous market, or of purchase decision-makers that with different roles interact with the product [4].

[3]The Five Ws (sometimes referred to as Five Ws and How, 5W1H, or Six Ws or: Who, what, why, how, where, when, with what.

[4]Journalists are trained to answer those five questions whenever they write an article or press release. The same kind of questions can also be the best way to start with inquiries in a consultancy meeting.

[5]*how* can also be covered by *what, when*, or *where*, but it is often a separate question because, sometimes, it proves to be the most interesting.

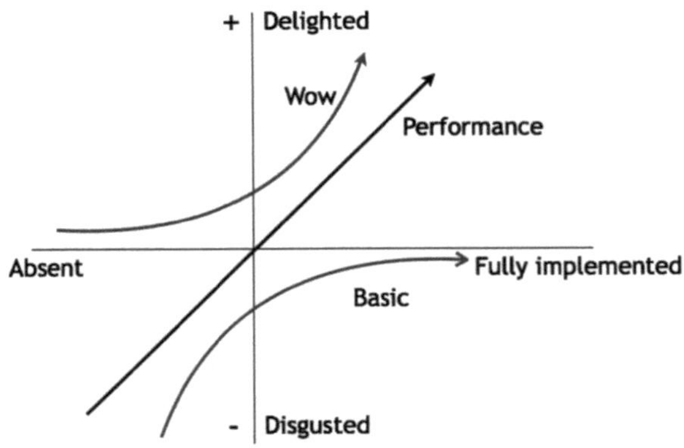

Fig. 3.1 Kano model for customer satisfaction

1. Expected needs that are generally unexpressed because the user implies that they exist.
2. Expressed requirements: specific characteristics that a customer requires explicitly from a product.
3. Unexpressed needs but latent: characteristics of the product that the user does not require specifically.
4. Exciting requirements: characteristics of the product that the customer does not expect and that make it possible to differentiate the product from the competition products.

A good design follows the hierarchy of needs. The lower-level needs must be satisfied before higher-level needs.[6] Needs of the lower level are *Functionality*, then moving towards higher level needs, *Reliability*, then *Usability, Proficiency and Creativity*. See [16].

3.2.1.1 Kano Model

A diagram that synthesizes the hierarchy of customers' needs and emphasizes that the perceived value of a design corresponds to its place in the hierarchy is given by a visual representation of customer satisfaction as a function of the implementation of special characteristics in the product design: the so-called Kano Model Fig. 3.1. It was developed in the 1980s by Professor Noriaki Kano, writer and consultant in the field of quality management [14, 24].

1. Some basic requirements are not usually mentioned unless they are absent from the product. In this case you would be disgusted. They are considered *Basic Requirements*. The points on the lower curve show this situation, at different levels of implementation.

[6]In order for a design to be successful, it must meet people's basic needs before it can attempt to satisfy higher-level needs [15].

2. Then a customer has expressed and unexpressed but latent needs that, if considered, provide an increase of satisfaction. They are called *Satisfiers*.
3. Finally, a customer has some *unspoken* and unexpected needs that, if satisfied (*wow requirements*), produce an added excitement: they are called *Delighters*.

Designers, marketing, and customer representatives are the main prompters of these suggestions. We must also considered the extreme case, that sometimes is verified, when the customer him/herself is the author of improving or extending the product use [10].

3.3 Logic Based Concept Selection

The first category of tools to carry out scientific research of the market is based on a logical treatment of data collected by customer interviews. This selection must be made in close collaboration between marketing and managers for business decisions and although it does not leave aside an internal activity carried out, e.g. in the company's laboratories, it is mainly focused on the market.

Affinity Diagram

A method that can be used to assess the relative importance of the needs and to clarify their internal relation, makes use of tools called affinity diagrams. The seminal work on this method is by Jiro Kawakita 1920–2009, a Japanese Anthropologist [19, 26, 29]. This method has an anthropological origin because it relates to the behavior of our mind. The affinity diagram is used to collect and put together similar statements and finally to discover causal *independent* relationships between requirements and needs. It is based on the *Logic* with the task of clarifying confused statements and separating requirements in such a way to avoid reciprocal influence and to guarantee independence between one and others. This Affinity process is formalized in *Affinity Diagrams* [8, 12]. It is a typical team activity, organized with the following steps:

1. *Recording*: Recording each idea of the individual team member on separate sticky notes or cards and show ideas on a large surface to all team members.
2. *Grouping*: Look for ideas that seem to be related in some way, and sort cards into groups of related needs until all cards have been used.
3. *Selecting Headings*: Once the cards are grouped, select a heading for each group.
4. *Combining* groups in *Super groups* in such a way to reduce requirements to essential (indispensable) ones.

The following example shows use of affinity diagrams for "disentangling" confused needs obtained by customers interviews.

Example 3.1 (*Needs analysis for the definition of motorcycle characteristics*) Customer interviews can produce confusing (i.e. interrelated) requirements, see example in Fig. 3.2 (Table 3.2).

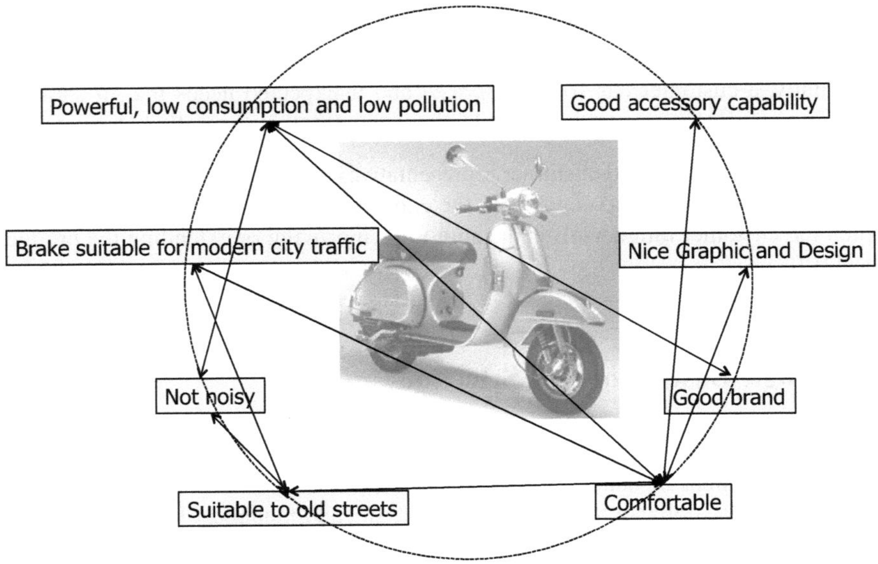

Fig. 3.2 Interrelated needs expressed by the customer on the design of a motorcycle

Table 3.2 Interview analysis of potential users

Interrelated needs	First affinity diagram	Actions on:
Nimble in traffic	Powerful and elastic	Engine power
Fitness for any road	Efficient suspension	Suspension
Warranty covering a long time	Type of warranty	Brand
Low consumption	Low consumption	Engine curve
Tolerance of heavy traffic	Efficient braking system	Breaks elasticity
Nice	Graphics, color, name	Design, graphics, colors
Extra objects	Compartment	Accessories

Table 3.3 Final seven headings of affinity analysis

Engineering	Design and imaging
Engine	Graphics
Brakes	Design
Suspension	Brand
	Accessory

Effects ↓ / Causes →	A	B	C	D	E	Total
A						0
B	9			9		18
C	1	9		1		11
D	9				3	12
E	9		9			18
Total	28	9	9	10	3	59

Most dipendent (B, Total 18); Most influencing (A, Total 28)

Fig. 3.3 The first use of the interrelationship matrix. Modified from [11]

	Effects / Causes	Engine	Brakes	Suspension	Graphics	Design	Brand	Accessory	Total	Percentage %
Engineeringing	Engine	1	2	2	2	0	2	2	11	22.4
	Brakes	0	1	0	2	0	2	2	7	14.3
	Suspension	0	2	1	2	0	2	2	9	18.4
Form and Imaging	Graphic	0	0	0	1	0	2	1	4	8.2
	Design	2	2	2	2	1	2	2	13	26.5
	Brand	0	0	0	0	0	1	1	2	4.1
	Accessory	0	0	0	1	0	1	1	3	6.1

Fig. 3.4 Interrelationship matrix for a motorbike design (Table 3.3). Modified from [11]

Interrelationship Matrix

The first use of an interrelationship matrix is to document the internal dependency among the requirements, variables that are indicated generically with symbols, Fig. 3.3.

Continuing the previous example we apply the analysis through the interrelationship matrix to the following *confusing* requirements, Fig. 3.4:

The assessment of the importance can be made with different metrics. The matrix in Fig. 3.4 shows the values of the total sums as a percentage of the total sum (=49).

In order to test the robustness of the analysis the interview was repeated with other individuals belonging to the same market segment.[7] The results of the mean values

[7] Young people aged 25–35 years of higher learning, living in different western countries.

	Affinity Diagrams Headings	Mean	Standard deviation
Engineering	1 Engine	10.04	2.14
	2 Brakes	8.8	2.12
	3 Suspension	7.92	2.05
Form and Image	4 Graphic	2.96	2.66
	5 Design	7.23	2.78
	6 Brand	6.00	3.44
	7 Accessory	4.42	1.96

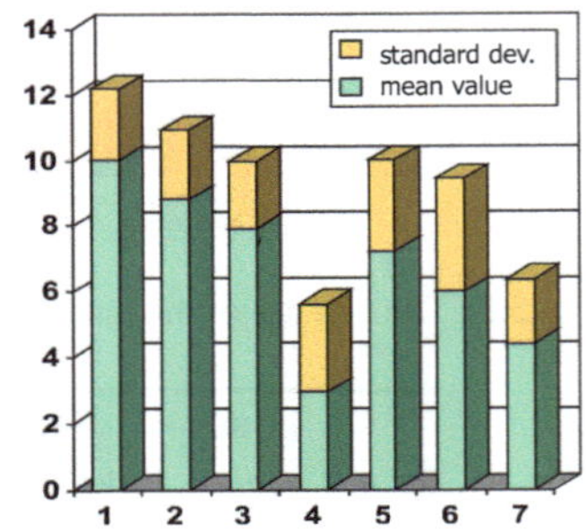

Fig. 3.5 Mean and standard deviation of the importance scores in repeated interviews of young people. Modified from [11]

and standard deviations of the importance scores are shown in the diagram on the right side of Fig. 3.5.

Analysis of Relative Importance of Customer's Requirements

The second use of the interrelationship matrix is to highlight the causal relationships of dependence and/or the relative importance of different requirements.[8] We thus define priorities and determine optimal sequences of actions.

The input data in an interrelation matrix are often defined by the needs coming from an affinity diagram. The implementation of the analysis consists of the following points:

- Formulate the matrix of requirements, entering the same items in rows and columns of a matrix.
- Compare one to one the importance of each requirement, using the following metric:
  - 1 same importance of row item and column item.
  - 0 row item less important than column item.
  - 2 row item more important than column item.
- The absolute and relative sum of each row gives the index of importance of each variable.

This tool can be used in different phases of the design and manufacturing process etc. every time the relative importance of items (not numerically expressed) is not

[8]We can speak of *requirements* but, at this stage, it would be better to speak of *general wishes* yet expressed in a qualitative way, because they derive directly from the *customer's voice*.

obvious. This can be interpreted as application of the Principle of ordination in Cartesian terms.

Relationship Matrix

A further logical tool is presented with the aim of interpreting the customer's voice and market trends. It is also based on the use of a matrix, that is called a relationship matrix. Through this tool it is possible to quantify the logical relationship between requirements, (Fig. 3.6 on the left) namely the WHAT (what the customer wants), and the HOW (how to meet the needs, by means of which technical choice).

WHATs in rows represent the customer's requirements to be met, while HOWs in columns are how to meet them, i.e. are the parameters (or the actions to be implemented), to be transformed into technical requirements, (Fig. 3.6 on the right).

- The coefficients of relationship are the following:
  1. No relationship = Empty box
  2. Weak relationship = 1
  3. Medium relationship = 3
  4. Strong relationship = 9
- The maximum value of the sum per row points out which requirement is more influenced by the column variables.
- The maximum value of the sum of the column points out which column parameter has the maximum effect on the row requirements.

If the relationship analysis is limited to the matrix on the left side of the Figure, this procedure of the requirements evaluation would not be so useful: in fact, it would not take account of the different importance of requirements.

WHAT \ HOW	Parameter /Action 1	Parameter /Action 2	Parameter /Action 3	Parameter/ Action 4
Requirement A		3	9	
Requirement B	9			
Requirement C		1	3	9
Requirement D	1		1	
Requirement E	3			3
Parameters Influence on Requirements	13	4	13	12

How / What	Impor -tance	How 1	How 2	How 3	How 4
Req. A	**10**		3	9	
Req. B	**10**	9			
Req. C	**6**		1	3	9
Req. D	**4**	1		1	
Req. E	**2**	3			3
Techn. Import.	absol.	100	36	112	60
Tech. Import. (Relative)	% of total	**32.5%**	**11.7%**	**36.4%**	**19.5%**

Fig. 3.6 Examples of relationship matrices. Modified from [11]

If, vice-versa, the requirements are classified according to the previous interrelationship matrix (or the previous objectives tree), we can introduce a column of *weight* of the relative importance of each requirement, (second column of the matrix on the right side). Then, the effective technical importance of each requirement *What* is obtained by the product of the relationship coefficient with the importance level, and the effectiveness of action (*How*) is evaluated by summing the column products.[9]

3.3.1 QFD

Quality Function Deployment (QFD) is the obvious conclusion of the previous analysis:

- 5W Rule.
- Affinity Diagram.
- Interrelationship Matrix.
- Relationship Matrix.

The method was developed in 1966 in Japan and gave significant support to the industrial development of that Country. According to the words of the inventor [1], QFD is:

> A method to transform qualitative user demands into quantitative parameters, to deploy the functions forming quality and methods for achieving the design quality into subsystems and component parts, and ultimately to specific elements of the manufacturing process.

The procedure for its use must be as simple as possible:

- It should not be considered a transient technique but it should be continuously applied in product development.
- It has to be integrated with other tools, such as: FMEA (Failure Modes and Effects Analysis), DOE etc.[10]
- It can be modified and/or supplemented to meet different needs: the concept of QFD is sufficiently robust to accept different forms of application.

---

[9] In this way the value of the weighted sum takes into account two aspects: the suitability of the technical choice in satisfying the row requirement and the importance of this requirement as perceived by the user. Thus sums in the last rows indicate the priority actions we should take to satisfy the most important requirements, with the most effective choice of technical parameters,

$$\sum_{i=A}^{E}[Imp \times R]$$

.

[10] See subsequent chapters.

In brief, the results that must be expected from QFD application are:

1. Documenting the environment where the product is used (in order to properly frame the expressed and unexpressed needs of potential users of the product).
2. Classifying needs in a logical order.
3. Assessing the relative importance of the needs.
4. The understanding of the customer needs is summarized in *product planning matrices or house of quality*. These matrices are used to translate higher level *what's* requirements or needs, issued from marketing, into lower level *how's* product technical characteristics selected to satisfy these needs. Requirements are desired, expressed in terms of *perceived quality*, while specifications are expressed in exact technical terms, challenging for the design, deployed and transmitted to all levels of the design as assembly groups, sub-assemblies, modules, etc.
5. Identifying technical requirements that have direct influence on needs satisfaction, in order to formulate full *technical specifications*.

Finally, QFD must be used as a nominal group technique.[11] The team members must come from different business functions, and maintain ties with their own department, but for the duration of the project depend on the group. They undertake to implement later what the group has decided.

3.3.1.1 The House of Quality

The house of quality is a sort of iconic summary (typical of oriental culture that illustrates the method with the shape of a house) based on the composition of matrices previously presented together with other bench-marking charts, Fig. 3.7. Figure synthesizes information on:

- User's WHATs requirements (#1) A, B, C, D, E, classified in order of decreasing importance from a max. rating of 10.
- HOWs actions (or parameters) selected to satisfy the requirements. The comparison is given through the relationship matrix.
- Directions of choices to improve or even innovate solutions.
- Benchmark with competitors solutions.[12]

[11] Nominal group technique (NGT) (meaning in name only, because verbal discussion is avoided) is a structured variation of a small-group discussion to reach consensus. It is a more-controlled variant of brainstorming used in problem solving sessions to encourage creative thinking, without group interaction. (Each member of the group writes down his/her ideas which are then discussed and prioritized one by one by the group). It is also called nominal group process [5].

[12] Benchmark: a standard or point of reference against which things may be compared or assessed.

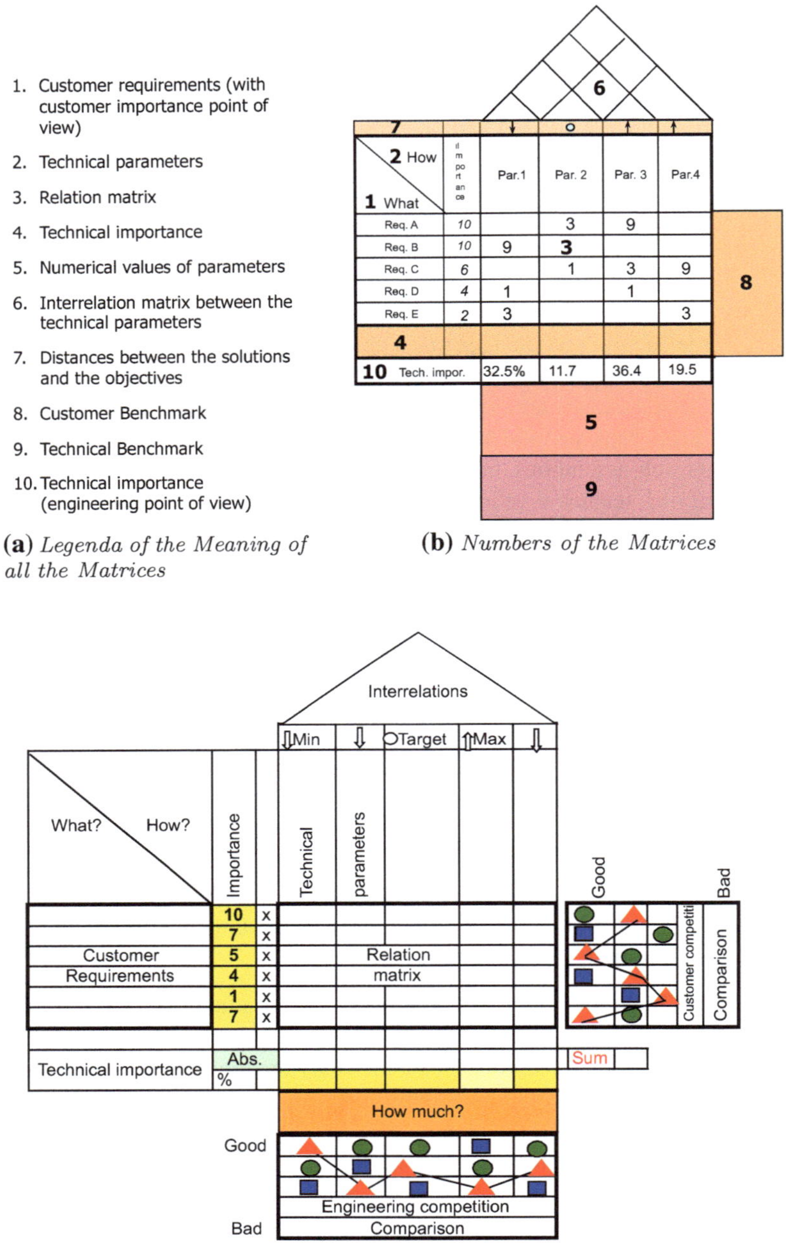

(a) *Legenda of the Meaning of all the Matrices*

(b) *Numbers of the Matrices*

(c) *Details and Meaning of the two Benchmark Matrices #8 and #9*

Fig. 3.7 The house of quality. Modified from [11]

3.3.1.2 The House of Quality "Boxes"

Customer requirements (What), Column #1
Technical Parameters (How), Row #2
Relationship matrix, #3

Technical Difficulty, Row #4 Below the relationship matrix a row gives the level of difficulty for carrying out the chosen solutions. (It gives a preliminary idea of technical difficulties to achieve a design solution, on the basis of current technical knowledge).

How Much, Row #5 The box under the main matrix is called HOW MUCH? that means: to what extent? In it, the numerical ratings of technical requirements are entered.

Roof of the House, Matrix #6 The roof of the house is an interrelationship matrix, (#6), Fig. 3.8, of the same type of matrix of Fig. 3.4. As the metric used in that one (0, 1, 2) was purely conventional, here too we can use a series of symbols that make sense of positive strong, positive weak, negative strong and negative weak interrelationships, with the following meanings:

- A positive relationship between two technical parameters means that an increase of a parameter value generates an increase of the others. This can be interpreted as potentiality for improvement.
- A negative interrelationship between two technical parameters means the increase of the value of one parameter generates a decrease of the value of another parameter: in this case we must give up this solution and devise a new one.

Objectives matrix #7 The row indicates, by means of a graphic symbol, if the HOWs and the corresponding quantified values are sufficient to meet (Symbol 0) or not meet, (arrow) the user's requirements (and, in this case, in what direction the values should be varied to bring them closer to the desired goals).

Customer benchmark, matrix #8 A comparison with the best competitors from the users' point of view.

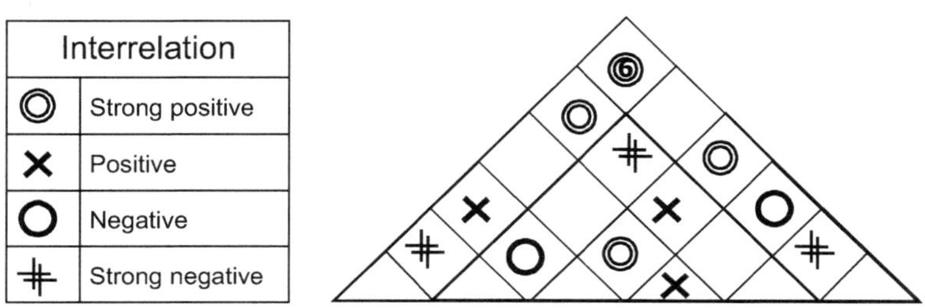

Fig. 3.8 Details of the house roof matrix. Modified from [11]

Technical importance, matrix #5 Numerical values.
Technical benchmark, matrix #9 A comparison with the best competitors from the engineering point of view.
Percent of Technical Importance, row #10 HOWs parameters are shown in percent form.

Note that the *Customers' perceived quality* can be in conflict with the quality offered by a purely technical decision. In such event it is necessary to review the technical solutions to understand the reasons of discrepancy, for instance by a design review.

The *Technical Importance* of each action takes into account both the importance of the judgment on the requirement (expressed by the customer/user), and the judgment on the relationship between this requirement and the technical action, (established by the engineering) to satisfy it.

3.3.2 *Examples*

Simple examples of the use of interrelationship and relationship matrices are shown on the following pages, with the aim of familiarizing with the QFD method.

Let's start with an interrelation matrix to assess the relative importance of a series of WHAT requirements, in the house of quality Fig. 3.9 and we shall continue with an example of a relationship matrix, marked with (#3), expressing a relationship between WHATs (#1) and HOWs (#2), for objects of common use.

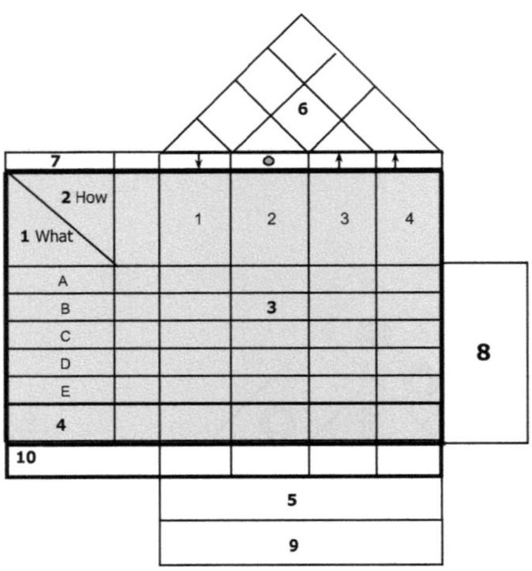

What \ How	Importance	Load or moment for opening	Load or moment for closing	Choice of new devices	Colors	Materials	
Ermetic closure	10	9	9	9			
Signal of break	7	9		9			
Reclosable	7		9	9	9		
Look nice	3			9	9	1	
Easy to open	3	9	9	9	3		
Healthy & Ecologic	3			9	1	9	
Absolute Techn. importance		180	180	297	102	30	789
Importance %		23	23	37	13	4	

Fig. 3.9 Analysis of relative importance of the user's requirements for a closure device of a bottle and of technical parameters that satisfy them, through the use of column, What (#1) and row, How (#2), and of a Relationship Matrix (#3). Modified from [11]

Example 3.2 (*Interrelationship and Relationship Matrices for a Closure Device for a Bottle*) This test of an introductory nature, is limited to the building of interrelationship and relationship matrices without design developing, just to show the evaluation method.

1. Who? General Use.
2. What? It must allow opening and re-closing.
3. Where? Without environment restriction.
4. When? No time restrictions.
5. Why? To allow repeated use.
6. How? It must be used without special tools.

The following Fig. 3.9 shows a comparison of an interrelationship matrix for establishing the relative importance among the requirements, and of a relationship matrix to show which technical parameters must be firstly considered in the following design development.

Example 3.3 (*Benchmark Matrix (#8) for the comparison of a car safety belt with competitors' products*) We want to clarify the design requirements of a safety belt for cars, picking up the needs of a potential user, drawing a first level of an affinity diagram, and the structure of a relationship matrix, imagining comparing the customer's point of view with the engineering perspective. In particular, we want to prepare the #8 matrix from the user's point of view, by means of a comparison with competitive products.

Customer Requirements

1. Who?: Persons of different height and weight (see anthropometric tables).
2. What?: It must offer safety in collision or rollover and be easy to open (in critical situations).
3. Where?: On cars of any power and size.
4. When?: Under normal travel conditions (not racing).
5. Why?: To provide greater security than existing solutions.
6. How?: It must suggest (force) use by a distracted user.

First Selection

- Good fitness for adults and children.
- It must restrain the body to the car seat and prevent the person's head from hitting the dashboard and the windshield and avoid other more serious damage to the body, up to a assigned body weight (anthropometric table), at a maximum given speed.
- It should not be a nuisance due to excess pressure on the body, abrasion of clothing, poor heat dispersion, or stiffness of the materials etc.
- It must be of easy to close, requiring no excessive twisting and bending of the body and clear and simple to open.
- It must not dirty clothing and be made from aesthetically pleasing material.
- It must not interfere with entering and exiting the vehicle.

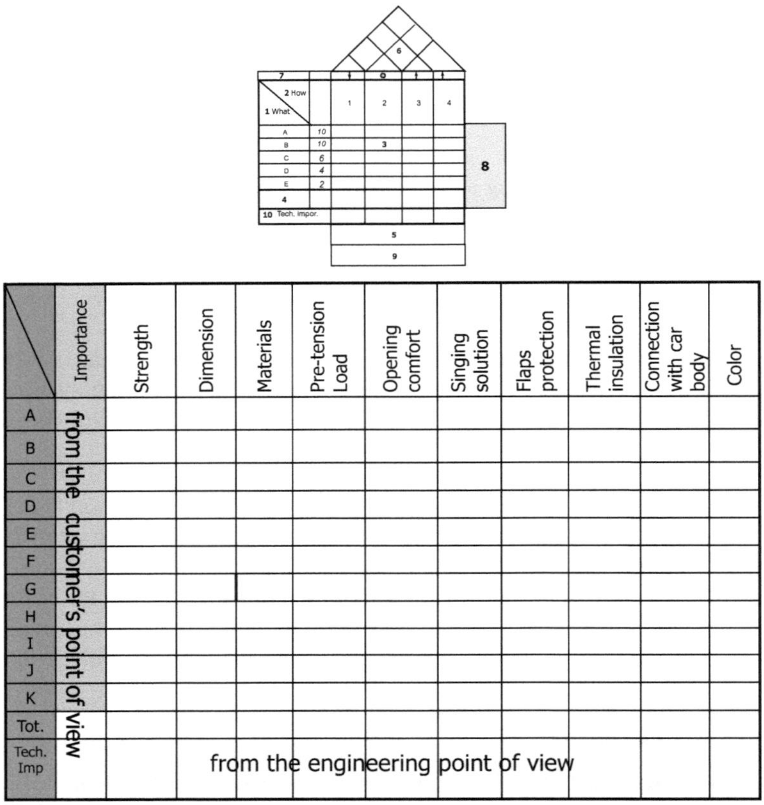

Fig. 3.10 A relationship matrix for QFD analysis of a safety belt. Modified from [11]

- Desired innovation elements (why?):
  - It must be suitable for pregnant women to use.
  - It must help distracted people.
  - The color must match the car's color.

First Affinity Diagram

1. Limited pressure on the body.
2. Not breaking under impact.
3. Not hurting in the case of collision.
4. Dissipating body heat.
5. Suitable for children.
6. Suitable for pregnant women.
7. Easy opening and closing.
8. Nice fabric.
9. Not easily getting dirty.

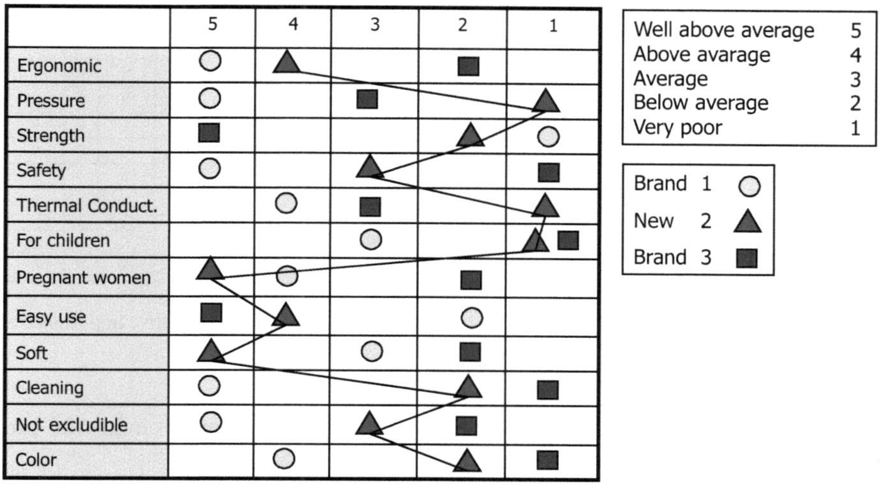

Fig. 3.11 Customer benchmark matrix #8: comparison with the best competitors from the users' point of view. Modified from [11]

10. Giving strong advice to its use.
11. Matching the car body's colors.

Figure 3.10 shows the type (not developed) of the relationship matrix and Fig. 3.11 the benchmark matrix.

3.4 Usability Based Concept Selection

Given the *maturity* of some products, the possible improvements are minimal and relate to innovative features that reduce the cost or introduce new useful features.

This observation suggests a slightly different approach that does not avoid the use of the methods previously exposed, but is predominantly focused on similar products on the market, to relieve strong and weak points (details) of existing solutions. We have called this the *experimental method*, because it takes into account the solutions already developed, with the intent of finding possible improvements, starting from the defects and strengths of the competitors.

An example of the design of a commonly used kitchen device (Juicer) is presented up to the development of the conceptual phase.

Example 3.4 (*Developing the product concept for a Juicer- Lemon/orange-squeezer*)

Interviews for Systematic Search

The following questions try to identify the product use-context through interviews with groups of potential users.

	1	2	3	4	5	6	7	Total	Norm
1 Usability	1	2	2	0	0	2	0	7	5
2 Transport	0	1	0	0	0	0	0	1	1
3 Stock-ability	0	2	1	0	0	2	0	5	4
4 Cleaness	2	2	2	1	2	2	2	13	10*
5 Design	2	2	2	0	1	0	2	9	7
6 Silent	0	2	0	0	2	1	0	5	4
7 Cost	2	2	2	0	0	2	1	9	7

How / What	Importan.		Weight	Effort	Mechatr.	Plastic	Dimens.
Cleanness	10	x	1		9	9	
Design	7	x	9		9	9	3
Stock-ability	5	x	9		1	9	9
Transportability	4	x	9		3	9	9
Usability	1	x		9	9		1
Cost	7	x	1	9	9	9	9
Technical Importance.			161	72	242	327	265
%			15	6	23	31	25

Fig. 3.12 Interrelationship and Relationship matrices. For documenting the relative importance of the items. Modified from [11]

1. Who?: Who is using it?: Domestic use, adults and children, not industrial use. Special attention to children's taste.
2. What?: What is the use?: for squeezing juice out of a fruit (lemon, orange, etc.).
3. Where?: Where is it used?: On a kitchen table with power network.
4. When?: When is it used?: Temporarily.
5. Why?: Why is it used?: For peel-resistant fruit.
6. How?: How is it used? It needs several hand movements.

First Selection Through an Affinity Diagram

The affinity diagram gave the following results[13]:

- Easy to use and with low driving effort.
- Easy to transport.
- Compact to store.
- Easy to clean.
- Good design and attractive colors.
- Silent.
- Inexpensive.

Selection Through Interrelationship and Relationship Matrices

Figure 3.12 shows the first two tools for documenting the *relative importance* of the user requirements and the *technical answers* (matrix # 3 of the House of Quality) to satisfy them.

Figure 3.13 shows the matrix of technical importance with the evaluation of the HOW MUCH parameters, (Matrices # 10 and # 9).

Usability Based Selection and The House of Quality

Figure 3.14 shows at the top the usability comparison of competitor products, while the House of Quality is shown at the bottom of the same Figure.

[13]We detected each requirement independent from each other and therefore a preliminary analysis of dependence between the requirements, performed with relationship matrices, is deemed unnecessary.

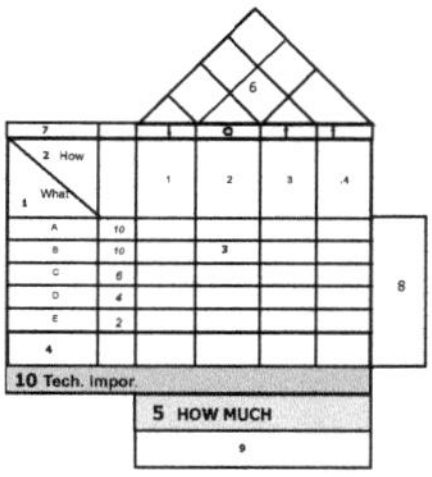

	2 Parameters (HOWs)				
	Weight kg	Effort	Mechatronic	Materials	Dimensions
10 Technical importance %	15%	7%	23%	31%	25%
5 HOW MUCH	2	5 N	Automatic switch and temperature control	Not breakable plastics	250x 250x250 mm

Fig. 3.13 The HOWs parameters classified according to importance. Modified from [11]

The QFD method at the first level (see Fig. 3.15 and the first line from the top) does not provide only a technical specifications list, though accurate and complete [21]. The method provides much more as task clarification, because it suggests the direction of the design improvement. It is not yet a presentation of conceptual solutions, but it puts the development on the right path and reduces errors. Together with the basic HOWs parameters described in #2 and evaluated by the matrix #3, we are able to:

1. measure customer expectations benchmarking usability (matrix #8) and engineering evaluation (matrices #4, #5, #9, #10);
2. estimate positive and negative interrelations between the different technical requirements (matrix #6), finding possibilities of improvements (matrix #7).

3.4.1 Conceptual Design Process Using QFD

LEVEL 1 of market research, The first house of quality QFD leads to the definition of technical requirements (and/or of actions required to meet them). According to the QFD inventor, the method should be used at successive levels of detail, in the all phases of the conceptual design process, Fig. 3.15:

LEVEL 2: Problem identification. In a second house of quality, WHATs inputs of a new relationship matrix are now the technical requirements, (or technical actions), identified by the previous application. WHATs are classified according to the technical importance, which reflects the priorities jointly assigned by users and technical experts. The HOWs outputs of the second house are now formed by the functional structures.

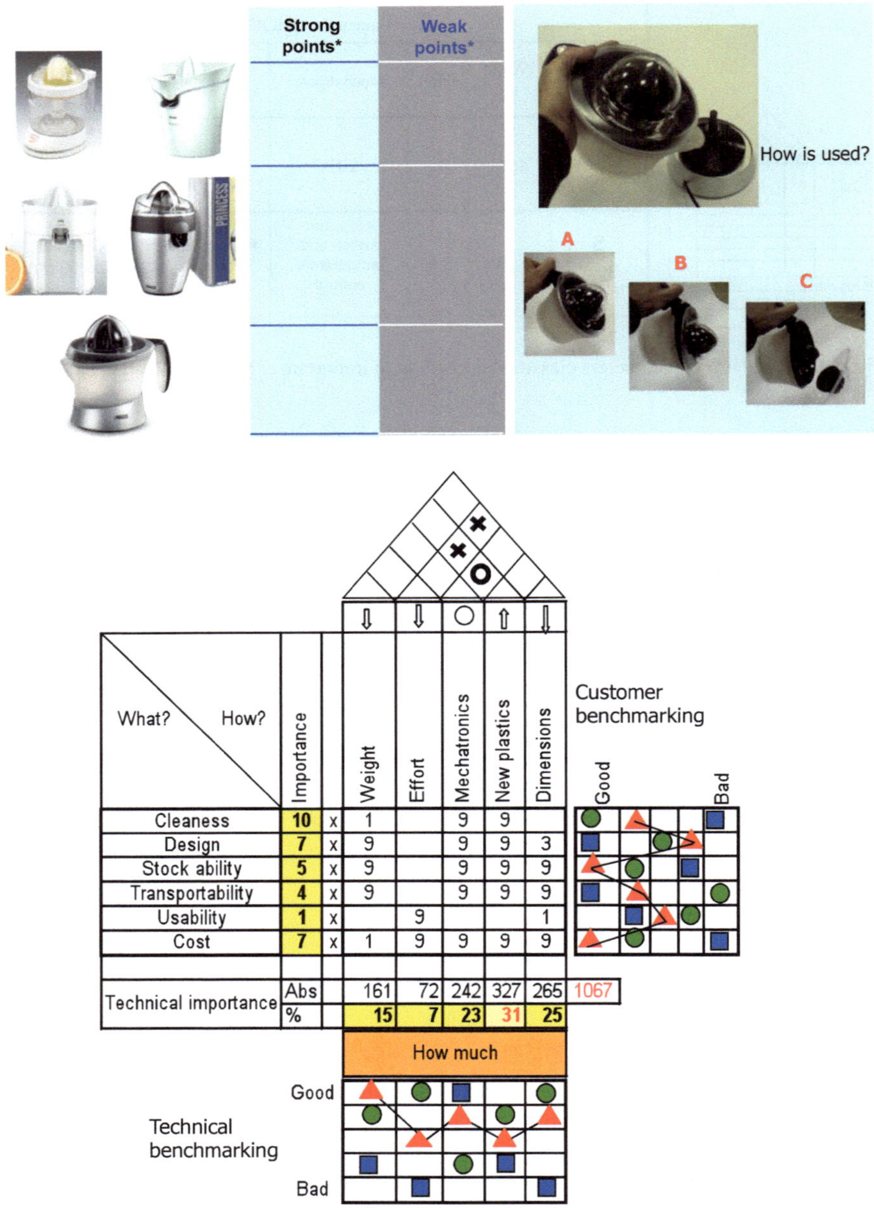

Fig. 3.14 Example of usability comparison of competitors products. The results are omitted and house of quality of the juicer design. Modified from [11]

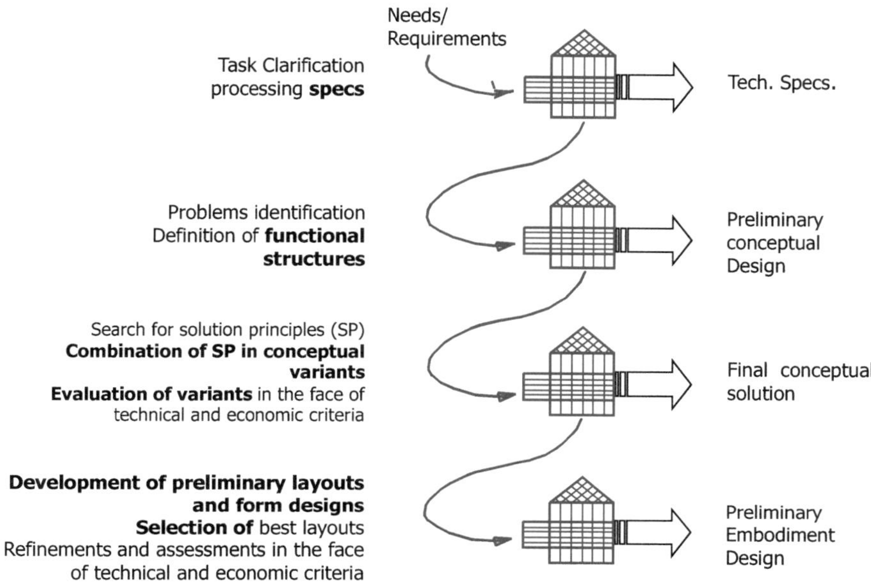

Fig. 3.15 Explanation of the House of Quality. Modified from [11]

LEVEL 3: Search for solution principles (SP). WHATs are the functions identified and classified in the previous step. Combinations of solution principles must now be found. Then solution principles are combined in conceptual variants. They constitute the outputs or the HOWs of the matrix. At this knowledge level it is important to make initial economic considerations on the costs of different conceptual variants, to complete in the next step.

LEVEL 4: Development of preliminary layouts and form design. The combinations of conceptual variants, identified in the previous step, are now WHATs inputs to the matrix of this fourth level. We look for the components necessary to a preliminary layout. They are the HOWs outputs and constitute the result of the conceptual design process up to a preliminary embodiment design.

3.5 Design Development

This is another important point of product planning. We can continue the previous example to develop this point. It is divided into the following phases:

- Functional Structure
- Morphological Matrix
- Industrial Design

Functional Structure

Figure 3.16a shows an example of functional structure of the first and of second level.

Morphological Matrix

Figure 3.16b shows the morphological matrix for the comparison between two variants.

Industrial Design

Figure 3.16c shows a prototype industrial design.

3.5.1 *Design Review*

In the case of a conflict between technical quality and user (perceived) quality a *Design Review* is necessary. A Design Review is a formal evaluation of a solution design, a control phase on something that is being developed but not yet completed. It must thus be planned before the beginning of a new phase and accompany any phase of the design development. The ultimate design review, if successful, triggers the product launch or product release.

Design review uses a formalized checklist [17].

The method is based on checklists with a series of questions that highlight the critical points of the planned and existing solutions [13]. Questions are structured as follows:

- What are the reasons for a better perception of the quality of competitors' products?
- Is the solution of the competitors cheaper? (cost analysis)
- How did competitors solve the problem?
- Must you reduce tolerances and therefore increase costs?
- Do competitors' solutions require fewer parts?
- Are competitors' solutions more *robust* compared to the disturbances in the manufacturing process and in the use?

(The last question will be better clarified by the introduction of another method as *Design of Experiment*).

Reviews should be objective, unbiased examinations by appropriately trained employees who were not directly involved in that design activity. The design team should keep a record of all such reviews and necessary actions that arise from them. Review results and actions arising from them should be documented in report form and signed by Management employees as complete and accurate, [2].

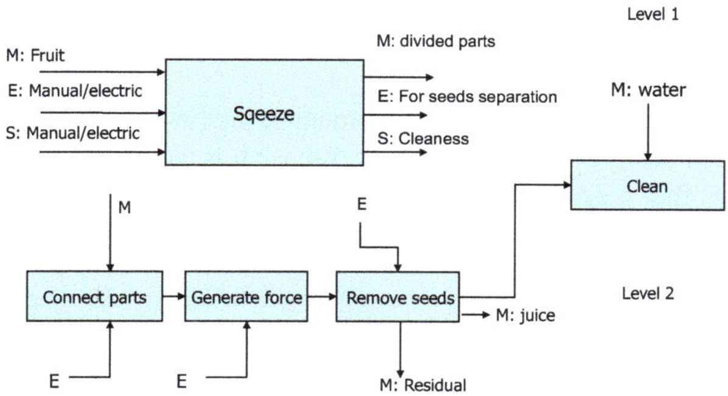

(a) *Functional structure of Level 1 and 2 of a Juicer*

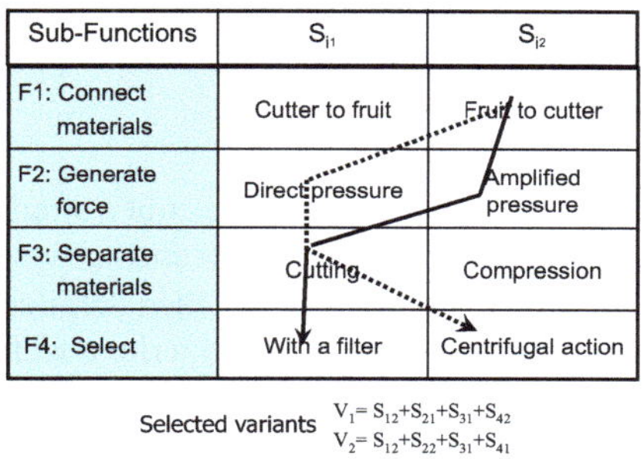

Sub-Functions	S_{i1}	S_{i2}
F1: Connect materials	Cutter to fruit	Fruit to cutter
F2: Generate force	Direct pressure	Amplified pressure
F3: Separate materials	Cutting	Compression
F4: Select	With a filter	Centrifugal action

Selected variants $V_1 = S_{12}+S_{21}+S_{31}+S_{42}$
$V_2 = S_{12}+S_{22}+S_{31}+S_{41}$

(b) *Morphological matrix of a Juicer Design for two selected conceptual variants*

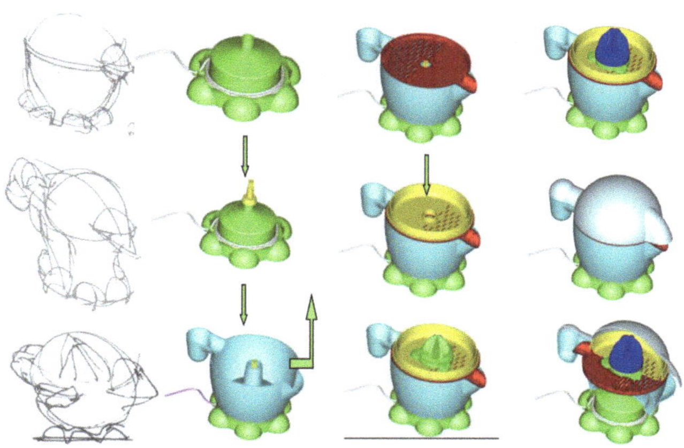

(c) *Prototype of a Juicer. K. Krsmanovic & G. Mijajlovic at a Course on Design at a Summer school of Bologna University, [11]*

Fig. 3.16 Conceptual Design of a domestic Juicer. Modified from [11]

3.6 Technical Specifications

The technical specification is the final document of the first phase of design planning, i.e. of conceptual design. In the most general sense it is a document that defines a set of requirements that a product (item, assembly, material, method, process, service, system, or work) must meet or exceed. Technical specifications shall contain, at least, the values of the following items:

1. Functional requirements.
2. Operating requirements.
3. Data on:
   - Energy.
   - Materials.
   - Information.
4. General constraints and limitations.

For complex systems, the technical specification is not a single document. It is composed of several documents, diversified for system and subsystems: drafting of a first synthetic specification defining the salient points, then detailed specifications follow. In the technical literature, guidelines exist to help write a technical specification for different types of products.

> The difference between the design of products for end-users (in a general market) and for customers in the industrial sector has already been recognized. For the former, market research, understanding and interpreting the *customer's voice*, and not only the functional values of the product (but also the semantic values), are essential for the definition of technical specifications, while for the second ones the design directly connects the requirements of clear-identified customers with the experts of specialized companies. In this case, the product (that can be a single prototype) is adapted to the specific customer requirements without the filter of interviews of several consumers.

3.6.1 Guidelines for Writing a Technical Specification

Guidelines and suggestions exist to help to write a technical specification. We suggest a model and a checklist applied to a previous example.

> A specification list must contain a detailed description of technical requirements, usually with specific acceptance criteria, stated in terms suitable to form the basis for the actual design development and production processes of an item having the qualities specified in the operational characteristics;...(a composite of the conditions, circumstances, and influences

that affect the item). From the Dictionary of Military and Associated Terms. US Department of Defense 2005, [9]. See also [3].

The first example of the contents of a specification list is the following [27]:

- Decide the acceptable dimensions and/or weight of the product.[14]
- Specify the full range of environmental (operational) conditions under which the product or assembly must meet the specification. If reduced performance of the product or assembly is acceptable at extreme temperatures and humidity, clearly note that in the specification.
- Place tolerances on the performance of the product or sub-assembly. Establish third party workmanship standards or safety standards that are to be applied to the product or assembly. This could include, for example, specifying that the product be CE certified.[15]
- Detail the technical specifications that the product or assembly must meet that are specific to that product or assembly. For example, an electronic assembly would have specifications for processing speed and product electronic interfaces, while a mechanical sub-assembly would have specifications for rigidity and load bearing capacity.
- Specify a life for the product or assembly. If scheduled maintenance or calibration of the product or assembly is acceptable, the specification must state that. The specification must state the acceptable conditions under which said maintenance or calibration of the product or assembly is performed and how often said maintenance or calibration is performed.

3.6.1.1 Example of Application of a Check List (From [18]) to a Previous Design

The following synthetic *Check List* is done for controlling the general properties and constraints (i.e. items that can not be forgot in every design).

It is a tool suggested in [18] and, according to the authors' opinion, deserves to be always applied, though some adjustments to different types of design are possible. Many of these *properties* are expressed by a suffix used for forming nouns of quality or capacity to be able to accomplish something. The checklist can be considered a means of avoiding the so called -ilities syndrome, i.e. a delayed check (when the design is too developed), of all these properties. We apply this checklist to the previous example (Table 3.4).

Example 3.5 (*Check List for the Juicer's Specs Control*)

We can see the example of application to the juicer design.

---

[14]The specification's items are expressed as *shall* statements.

[15]The CE Mark. The European Commission describes the CE mark as a *passport* that allows manufacturers to circulate industrial products freely within the internal market of the EU. The CE mark certifies that the products have met EU health, safety and environmental requirements that ensure consumer and workplace safety. All manufacturers in the EU and abroad must affix the CE mark to those products covered by the *New Approach* directives in order to market their products in Europe. Once a product receives the CE mark, it can be marketed throughout the EU without undergoing further product modification.

Table 3.4 Check list for a juicer design

-ilities	Adopted solutions
Safety	Electrical protection according to standards
Ergonomic and form design	Very easy to use (and to clean). It must meet children's taste
Producibility	Suitable for a small company
Reliability and controllability	Data on MTTF and on production controls
Assemblability	No. of components $\leq \ldots$
Transportability	Wmax: 2 kg
Usability	Not only for oranges
Comfort	With a noise $\leq$ 40 db
Maintainability	No maintenance for 5 years
Sustainability	% of Recycling
Cost	max: 80 €
Time	max... months

3.7 What We Have Learned from This Chapter

Design is not a single idea generation but a process, i.e. a systematic series of actions directed to some end. The transition from customer needs to technical specifications is the main point and is described by examples. They serve to clarify the task of the designer in respect of data that come from the environment (market, product context, civil society etc.) and from the potential user/customer. The proposed tools for a systematic market research have origin from the logic and from experience. For this reason we spoke of logic based and usability based concepts selection. QFD, is a tool that can manage both approaches in a synthetic frame, called House of Quality. So we have two possibilities to examine the reality: the first is based on interviews, the other on direct analysis of existing products. The task is to obtain clear and distinct requirements, i.e. independent one from another to which appropriate technical solutions can be given.

The design process is iterative in nature, i.e. based on the trial and error procedure. So we cannot think of a sort of algorithmic procedure, but, at the same time, we cannot shy away from the huge number of standards and guidelines that are a great source of design knowledge. Suggestions for writing technical specification are then given. As a general conclusion we should reflect on these points:

- The design process is not necessarily sequential in all phases.
- It is the problem-solving type, open to a variety of choices.
- Different approaches are admitted. The final aim is reaching the best conceptual variant as a starting point for any subsequent embodiment and manufacturing development.

References

1. Akao Y (1994) Development history of quality function deployment. In: The customer driven approach to quality planning and deployment. Minato, Tokyo 107 Japan Asian Productivity Organization, ISBN: 92-833-1121-3
2. Bizmanualz (2017) Design review checklist ISO template QP1100-1. https://www.bizmanualz.com
3. Business-Dictionary (2016) Definition of technical specifications. http://www.businessdictionary.com/definition/specification-spec.html
4. Business-Dictionary (2016) Market segment. http://www.businessdictionary.com/definition/market-segment.html
5. Business Dictionary (2016) Nominal group technique. http://www.businessdictionary.com/definition/nominal-group-technique
6. Business-Dictionary (2016) Product planning. http://www.businessdictionary.com/definition/product-planning.html
7. Consultantsmind (2014) 7 Key questions: who, what, why, when, where, how, how much? www.consultantsmind.com/../7-key-questions-who-what-why-when-where-how-how-much
8. Crow K (2016) DRM associates 2613 via Olivera, Palos Verdes, CA 90274, USA. http://www.npd-solutions.com/worksheet.html
9. Dictionary F (2005) Technical specification US DoD. http://www.thefreedictionary.com/technical+specification
10. Faulkner P, Runde J (2009) On the identity of technological objects and user innovations in function. Acad Manag Rev 34(3):442462
11. Freddi A (2004) Imparare a progettare, vol 1, 1st edn. Pitagora Editrice Bologna
12. Frigon NL, Mathews D (1997) Practical guide to experimental design, vol 1, 1st edn. J. Wiley & Sons, Inc
13. Gewirtz CD (1993) Developing new products with TQM. McGraw-Hill, ISBN-13: 978-0070235731
14. Kano N (5–10 May, 2011) Seminars at economy faculty- Rome III Univ. http://www.kanomodel.com/
15. Lidwell W, Holden K, Butler J (2003) Universal principles of design. Rockport Publication, l
16. Maslow A (1987–1954) Motivation and personality. Addison-Wesley
17. Minnesota University Service (2017) Design review checklist. https://www.uservices.umn.edu/sites/uservices.umn.edu/.../checklist_design_review.doc
18. Pahl G, Beitz W, Feldhusen J, Grote KH (2007) Engineering design: a systematic approach, 3rd edn. Springer, London
19. Plain C (2007) Build an affinity for KJ method. Qual Prog 40(3):88
20. Porter J (2010) Five Ws and One H: the secret to complete news stories. http://www.blog.journalistics.com/2010/five-ws-one-h/
21. ReVelle JB, Moran JW, Cox CA (1998) The QFD handbook. Wiley & Sons Inc., ISBN-13: 978-0471173816, ISBN-10: 0471173819
22. Schilling MA (2013) Strategic management of technological innovation, 4th edn. International Edition, McGraw-Hill Education. ISBN-13: 978-0078029233
23. Start-up Blender (2012) The difference between users and customers. http://www.startupblender.com/the-difference-between-users-and-customers
24. Stroud JD (2016) iSixSigma 2010–2016; The Kano analysis: customer needs are ever changing. http://www.isixsigma.com/tools-templates/kano-analysis/kano-analysis-customer-needs-are-ever-changing/
25. Suttle R (2017) Definition of product planning. http://smallbusiness.chron.com/definition-product-planning-5052.html
26. Wikipedia (2015) Affinity diagram. https://en.wikipedia.org/wiki/Affinity_diagram
27. Wikipedia (2015) Write a technical specification. www.wikihow.com/Write-a-Technical-Specification

28. Wikipedia (2016) Five Ws. https://en.wikipedia.org/wiki/Five_Ws
29. Wikipedia (2016) Seven management and planning tools. https://en.wikipedia.org/wiki/Seven_Management_and_Planning_Tools

Chapter 4
Invention and Innovation

Abstract Up to this point we met design methods with a predominantly technical approach, according to which the design process is developed as a sequence of actions related to the invention of solution principles. Through the previous approach we learned how to perform certain functions, i.e. with the invention of functional variants and also in certain cases, the construction of prototypes. However, especially design guidelines and the ISO 9001 standards now open up a broader scenario, in which a technical design is oriented to an industrial mass production. In this case, the problem-solving choices are no longer only a problem of technical invention. Mass-production requires the study of another coordinate of the problem: the design development within an industrial company. It concerns the "innovation concept" that is the core value played primarily within company strategies.

4.1 Innovation Versus Invention

In the everyday language, *invention* is confused with *innovation*. In reality, the two concepts are totally different. We can compare two definitions of invention and innovation (given by the same dictionary), in order to observe differences and analogies. About invention [3]:

> Invention is new scientific or technical idea, and the means of its embodiment or accomplishment. To be patentable, an invention must be novel, have utility, and be non-obvious. To be called an invention, an idea only needs to be proven as workable. But to be called an innovation, it must also be replicable at an economical cost, and must satisfy a specific need. That why only a few inventions lead to innovations because not all of them are economically feasible.

Conversely, according to the same dictionary [2], innovation means:

> The process of translating an idea or invention into a good or service that creates value or for which customers will pay. To be called an innovation, an idea must be replicable at an economical cost and must satisfy a specific need. Innovation involves deliberate application of information, imagination and initiative in deriving greater or different values from resources, and includes all processes by which new ideas are generated and converted into

A. Freddi and M. Salmon, *Design Principles and Methodologies*, Springer Tracts in Mechanical Engineering, https://doi.org/10.1007/978-3-319-95342-7_4

useful products. In business, innovation often results when ideas are applied by the company in order to further satisfy the needs and expectations of the customers.

Innovation could occur also if the designer improves or makes a significant contribution to an existing product, process or service. Innovation differs from invention because innovation presumes the existence of a final product, then of an entity able to produce it, then a production process and a service process. All this can only be achieved by an organized structure i.e. by an industrial process.

Invention is therefore a useful idea that is generated also by an individual who is identified and protected as the inventor, while innovation can be suggested by any single person but, in any case, is the result of a community activity of industrial type, as the process of growing ideas into practical use.

An innovation does not necessarily need invention. We find success in innovating something that already exists, such as providing elements of innovations in existing product or/and in existing production process or/and in service. Examples can be found of product innovation obtained by taking *off-the-shelf items*[1] [4, 5], and putting them together in a way never thought before and which satisfies new needs.

Even without inventions, a company can still come out ahead.

Example 4.1 (*Innovation without invention*)

Apple deeply innovated the mobile phone connecting together existing functions of calling, taking pictures and recording music and completely rethinking how they could be merged and efficiently used. The Apple watch represents an invention due to new interfaces assembled inside, but it becomes an innovation when it enters the Apple iPhone applications. The same happened in Japan for cars or TVs, that were innovated within world-leading companies and economies, without breakthrough inventions.

Probably the famous maxim of Thomas Alva Edison clarifies the distinction between invention and innovation:

None of my inventions came by accident. I see a worthwhile need to be met and I make trial after trial until it comes. What it boils down to, is one per cent inspiration and ninety-nine per cent perspiration.

Definition of invention is the concept of a new idea, a first step in a long process of meeting a worthwhile need. The ninety-nine per cent of hard work could be considered the passage from invention to innovation, that is really what drives economic growth.

At the beginning of the industrial revolution the two concepts of invention and innovation were hardly distinct, since the roles of inventor and entrepreneur were joined:

---

[1]Off-the-shelf item is commercially produced, ready-made, standardized, and regularly available equipment, goods, parts, software, etc., i.e. a product that is available immediately and does not need to be specially made to suit a particular purpose. Another related term is commercial off-the-shelf (COTS): an item that is commercially available, leased, licensed, or sold to the general public and which requires no special modification or maintenance over its life cycle.

> [....] I shall only observe, therefore, that the invention of all those machines [..], seems to have been originally owing to the division of labour. [..] Men are much more likely to discover easier and readier methods of attaining any object when the whole attention of their minds is directed towards that single object than when it is dissipated among a great variety of things. But in consequence of the division of labour, the whole of every mans attention comes naturally to be directed towards some one very simple object. It is naturally to be expected, therefore, that *some one or other of those who are employed in each particular branch of labour* should soon find out easier and readier methods of performing their own particular work, wherever the nature of it admits of such improvement. [...] All the improvements in machinery however, have by no means been the inventions of those who had occasion to use the machines.
>
> Many improvements have been made by the ingenuity of the makers of the machines, *when to make them became the business of a peculiar trade*, and some by that of those who are called *philosophers or men of speculation, whose trade it is not to do anything, but to observe everything; and who, upon that account, are often capable of combining together the powers of the most distant and dissimilar objects. In the progress of society, philosophy or speculation becomes, like every other employment, the principal or sole trade and occupation of a particular class of citizens.*
>
> (Adam Smith: The Wealth of Nations, Book 1, Chapter, 1776.)

4.2 Innovation of Product and Process

According to the Product Development and Management Association (PDMA), the act of *innovation [...] includes utilization of some inventions (i.e. the creation of a product or introduction of a process for the first time) as well as the work required to bring an idea or concept into final form* [20].

Schumpeter (1934), one of the most influential economists of the 20th century, identified innovation as an engine of business development.

Innovation's value to an organization manifests itself in different ways for both categories of process and product innovation [16]:

- Process innovations:
  1. Introduction of a new method of production.
  2. Developing new source of supply for raw materials or other inputs.
- Product innovations:
  1. Introduction of new products.
  2. Opening new markets.
  3. Creation of new market structures in an industry.

Other influential contributions to innovation come from the Harvard Business School. It has developed a variety of strategy concepts to bear on many of the problems facing corporations, economies and societies [18, 19].

We must recognize that innovation is often an abused word to assist companies in surviving the market difficulties. Many companies had developed innovation strategies as a reaction to the economic downturn, in order to be competitive in the market, and to get higher profit margins. The social dimension of innovation is proved by the fact that one can legitimately speak not only of a company's, but also of a nation's competitiveness, in close dependence on the capacity of its industries to innovate and upgrade.

The next cases are examples of industrial innovation. Product innovation is a competitive tool for the big economic blocks. Nevertheless, we must remember that innovation does not mean necessarily new products: today the industrial sectors rarely achieve superiority through the production alone.

Many disciplines may influence innovation. Schematically we can state that:

- Researcher = updates on the frontiers of technical science.
- Inventor = suggests new ideas.
- Innovator = sets up applications.
- Manager = organizes the process.
- Entrepreneur = finances the process and assumes the risks.
- Direct User = uses and benefits from the use.
- Suppliers = offer components, materials, innovative services.
- Other disciplines are also involved in the frontiers of innovation for various reasons: legal, public, financial, etc.

Items and constraints related with innovation can be summarized in the following Table 4.1.

Table 4.1 Product design environment

Social environment	Internal environment	Manufacturing
Physical environment	Company policies	Industrial design
Customers	Documentation	Ergonomics
Competitors	Patents	Delivery time
Safety	Fixed costs	Maintenance
Purchases	Variable costs	Disposal
Energy costs	Dimension	Life of service
Market constraints	Quantity	

Table 4.2 Innovation strategies

Intensity	Application area
Radical innovation design	*Products*
Incremental innovation design:	*Production process*:
1. Pioneer (best in class)	1. On the methods
2. Fast follower	2. On the organization
	Company organization
	Financial organization

4.2.1 Innovation Strategies

Several strategies can be implemented to innovate [13]. These strategies refer not only on the products but also on application areas. They identify resources, processes, risks, partners and suppliers, with reference to two main parameters: *Intensity* and *Application area*, Table 4.2, with the following explanation:

1. **Innovation in Intensity**
   Radical innovation design means new products for new markets [10]. It occurs when innovation derives from introduction of a new technology that creates new industries and transforms the way of living and working, (it is very rare). It can be regarded as a *transformational innovation* or breakthrough innovation, [6, 26], i.e. introduction to the market of new products that usually come from revolutionary changes,[2]
   Incremental innovation design. It involves incremental changes to existing products [10, 25] designs that are focused line changes or improvements in a firms existing product offerings[3]

   a. Pioneer (Best in class).
   b. Fast follower innovation design,. It focuses on improving existing technologies,[4]

2. **Innovation in Application Area**
   On the product:

   a. On the form, (e.g. new design of lamps and other interior items).
   b. On design, (e.g. development of new disk brakes for vehicles).
   c. On the function, (e.g. design of a new hard disk for computers).

[2]Radio, transistor, laser applications, new materials applications for airplanes as Boeing 787, new led bulb, totally new cars, liquid crystal displays (LCD), computers, portable phone etc.

[3]Lamborghini sport cars, Boeing 767 plane, etc.

[4]Examples are: FCA-Croma car, etc.

On the production process:

a. On the methods, (e.g. introduction of Laser for special operations, of Software for Management, etc).
b. On the process organization, (e.g. introduction of new methodologies as Computer Integrated Manufacturing (CIM), Just in time (JIT), Material Requirements Planning (MRP), Group Technology (GT), Concurrent Engineering (CE), etc).

On the company organization
On the financial organization

Consequently, leaderships and strategies arise from the choices of the type of innovation that we wish to introduce:

- Product leadership.
- Service leadership.
- Fast Follower, Lowest Price.
- Lean Production.
- Downsizing.
- Outsourcing.
- Shift from manufacturing to services.

4.2.1.1 Product Leadership

It is a competitive strategy which aims to build a culture that continuously brings excellent products to market.

Typically, companies of this type adopt tools as Supply Network Planning (SNP) to model the entire supply network for planning the flow of material along the production chain. SNP allows one to create feasible plans for purchasing, manufacturing, inventory, and transportation, and to closely match supply and demand. Companies that focus on product leadership push hard on product development. They create products that change the way the customer lives.

Examples of companies that have chosen the strategy of product leadership are, e.g. Apple, Sony, BMW, Ferrari, Lamborghini, Ducati etc.

4.2.1.2 Service Leadership

Companies that focus on service leadership push for a problems-free service in the all phases of the product life, as:

- Pre-Sale, i.e. a sale held in advance for select customers.
- Co-Design, i.e. a design shared with users and suppliers that become part of the design process and is joined with other experts in product development.
- Sales and market analysis.
- Sales together with bureaucratic and financial help.

- Assistance at use.
- Fast delivery.
- Assistance at Start-up phase.
- Maintenance with repairs and availability of spare parts.
- Follow up with retrofitting.
- Repurchase of obsolete products.

Examples are the following: Oracle, SAP, McDonalds, Genesys, Telecommunications Laboratories. All these companies developed operational excellence as strategic priority.

4.2.1.3 Fast Follower and Lowest Price

In the first case the company looks to see what the other competitors are doing, then does it better and cheaper, while in the second case it competes, basing its strategy only on price. Of course the company must be able to produce at lower cost increasing, at the same time, service, customization and sales network.

Examples of a fast follower strategy today is a company like Samsung. A comparison between Ford and GM is one of the most quoted historical case on lower price strategy.[5]

Today companies as Sharp, Acer, Toshiba can be considered fast followers after Apple that moved first with its products.

4.2.1.4 Lean Production

Lean principles are derived from the Japanese manufacturing industry. The company's design is oriented to maximize customer value while minimizing waste. Called also just-in-time production, lean production is a strategy related to an assembly line methodology developed originally for the manufacturing of cars.

Engineer Taiichi Ohno is credited with developing the principles of lean production after World War II [21].[6] Lean production principles are referred not only to

---

[5]Ford, with only one car model, focused on cost reductions gaining a large market share, while General Motors, in the first years, having differentiated product line, had less market share.

[6]*His philosophy is focused on eliminating waste and empowering workers, reduced inventory and improved productivity. Instead of maintaining resources in anticipation of sales of what might be required for future manufacturing, as Henry Ford did with his production line, the management team at Toyota built partnerships with suppliers. In effect, under the direction of Engineer Ohno, Toyota automobiles became made-to-order. By maximizing the use of multi-skilled employees, the company was able to flatten their management structure and focus resources in a flexible manner. Because the company was able make changes quickly, they were often able to respond faster to market demands than their competitors could. Many industries,(Nike, Caterpillar, Intel, Textron, John Deere, Ford, Toyota), including some companies for software development, have adopted the principles of lean production..*

manufacturing but also to a lean management. For a definition of lean-production, see [17]. Main rules of lean production are [21]:

1. Meet customer requirements.
2. Eliminate waste.
3. Minimize inventory.
4. Pull production from customer demand.
5. Do it right the first time.
6. Partner with suppliers.
7. Reduce the Time to Market.
8. Establish a culture of continuous improvement.

4.2.1.5 Downsizing

If originally downsizing was focused on reducing the number of employees maintaining the mission (through several techniques, e.g. to take early retirement, providing incentives), today downsizing has become an accepted set of strategies to reduce company size when it is necessary to transform products or services by eliminating unnecessary functions or decreasing the production of parts (modifying e.g. the ratio of Make-Or-Buy), or increasing cooperation with with related companies (e.g. through Co-design for sale or production).

Often, downsizing is characterized by massive layoffs, trying to save a struggling company. Many major corporations have downsized more than once in an effort to cut costs and remain competitive.

Example 4.2 (*Downsizing strategy for the production of internal parts*)
Figure 4.1 shows an example of application of downsizing strategy for the production of plastic moulding dies.

4.2.1.6 A New Business Model: Outsourcing and Shift from Manufacturing to Services

A modern economy, in the past few decades, has experienced a high rate of service economy in most developed countries. This tendency to dematerialization has brought some benefits as:

- Hire the good.
- Pay for Use: we provide the machine and they pay for the service.
- Profit on materials: we provides the product and they sell materials.
- Profit on the operating: we sale the product together with operating.
- Profit on services: we sell the product and its maintenance.

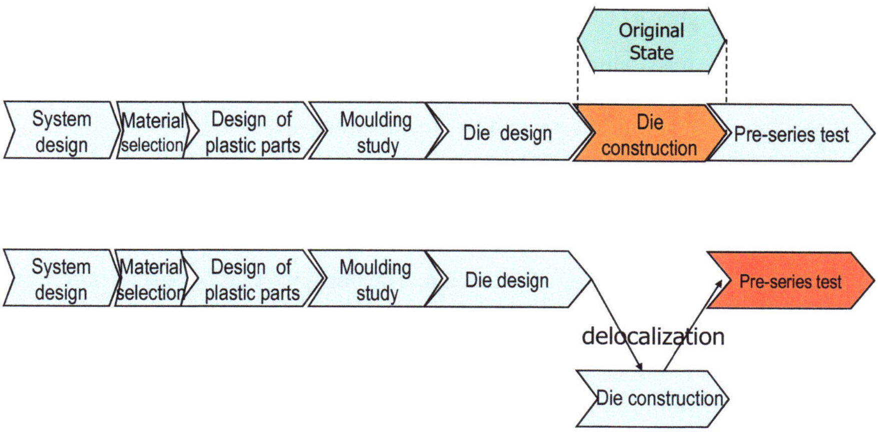

Fig. 4.1 Downsizing strategy decreases the production of internal parts

- Profit on financing: to introduce form of project financing.
  Build: to establish the infrastructure for business development, among which are people, offices, facilities, etc.
  Operate: to establish the offshore organization, i.e. to provide customer services to expand the market, acquire new customers, and gain more market share and revenue.
  Transfer: once the market becomes mature and the products and services from the vendor have been widely accepted by local market and customers, the vendor might consider terminating the BOT[7] relationship and make its own presence in the offshore country, at a proper time.

Note: A new trend of the outsourcing model is in the risk management that becomes, with this strategy, an important element of quality. It must be extended to "externally provided products or services" [12]. The reason is that outsourcing is subject to risk management and the organizations must address risk management in their evaluation of suppliers that are used for outsourcing.

4.3 The Life-Cycle of the Product and Designer's Tasks

The product life cycle describes the stages of a product from when it was conceptualized until it finally is removed from the market.

---

[7]Build-Operate-Transfer (BOT), is a form of project financing, whereby a private entity receives a franchise from the private or public sector organization to finance, design, construct, and operate a facility for a specified period of time. It is a business model which has the potential to offset project risks, while minimizing upfront investment and set-up time and offering wider service [9].

The main designer's tasks during the product development stages are:

1. *Introduction*: researching, developing and then launching the product. The designer's task is to monitor the product in its earliest forms of use. Sometimes it requires a fine tuning as a result of this first control.
2. *Growth*: when sales are increasing at their fastest rate, the designer's task is to adapt production to market demands and give support to customers to maintain customer satisfaction.
3. *Maturity*: sales are near their highest, but the rate of growth is slowing down, e.g. new competitors in market or saturation. The designer's task consists in refining and improving the product to maximize the satisfaction of the purchaser. It is time to start developing the new generation of the product.
4. *Decline*: This is the final part of the life cycle. Sales begin to fall. Task of the designer is to minimize the costs of maintenance and to develop transition strategies to provide the customers a new product.

Not all products reach this final stage. Some continue to grow, following choices of marketing strategies [14]. Special strategies can extend the life of the product before it goes into decline. They need to use marketing techniques such as [7]:

- Increasing advertising.
- Reducing price.
- Adding new features.
- Exploring new markets.
- Offering a new packaging.

4.4 Profit Versus Innovation

Innovation and profit do not follow a linear-proportional trend.

1. Low Level of Innovation: minimal changes to existing products. Examples at this level would be changing only the color of a product or putting it in a new package. Profits do not increase considerably. Simple re-styling of car body can be considered another example. It is a way to keep the market share, not to increment it.
2. Medium Level: vice-versa, re-styling of a car often foresees new features integrated into existing models or new differentiated versions of the same product to fit various types of customers. In this case, profit can considerably increase, with very low risk.
3. High Level: the highest level of innovation is where companies are able to create a breakthrough or revolutionary innovation that changes people' lives. Innovation of this type usually means high risks and longer term profits.
   Only few companies are equipped for achieving success at high innovation level.

The highest level of risk accompanies the highest level of innovation. That is the case of totally new products and systems that customers themselves do not know what they want.

The innovation level is not a constant for a company. The choice of the innovation level is a *moving target* that changes with time.

4.5 The Time Factor

Time is another critical factor for innovations. We can distinguish two parameters in the development process: Time-to-Market and Time-to-Volume.

Time-to-Market is defined as the length of time from product idea to the finished product.[8]

Time-to-Volume is a new measure of product innovation and manufacturing. This is considered preferable to *time to market*, as it deals with the time period required to get a new product into volume production [1, 15]. Said in other words, Time-to-Volume is the average time to develop a product from concept to launch and realization of commercially relevant business volumes [22]. It was observed that, reducing time to market can cause a tendency to shortcut tests necessary to verify manufacturing scalability. This can increase time to volume. Time to volume is then a parameter which better than time to market, takes into account all the uncertainties connected to the volume manufacturing, as, e.g. the engineering change orders, technology maturity, parts availability and testing.

4.6 Technology Readiness Level

In order to measure consistently and systematically the maturity of a particular technology, in comparison to the maturity between different types of technology, the Technology Readiness Level (TRL) was introduced originally by NASA[9] to support space technologies planning during the decade 1970–80, Fig. 4.2 [23].

It includes the 9 levels (NASA version) shown in Fig. 4.2. Today it is a method for demonstrating technology capabilities of promoting technological change and innovation and from this point of view it can be useful for any company on the market. In a time of commercialization of research results, it is a necessary method to communicate the readiness of technology from the technological research field to industry and allows a company to evaluate the readiness level of a certain technology,

[8]When the product is completed as to manufacturing but not yet sold or distributed to the end-user.

[9]NASA is abbreviation for the National Aeronautics and Space Administration, an independent agency of the executive branch of the United States federal government. DoD is abbreviation for the United States Department of Defense.

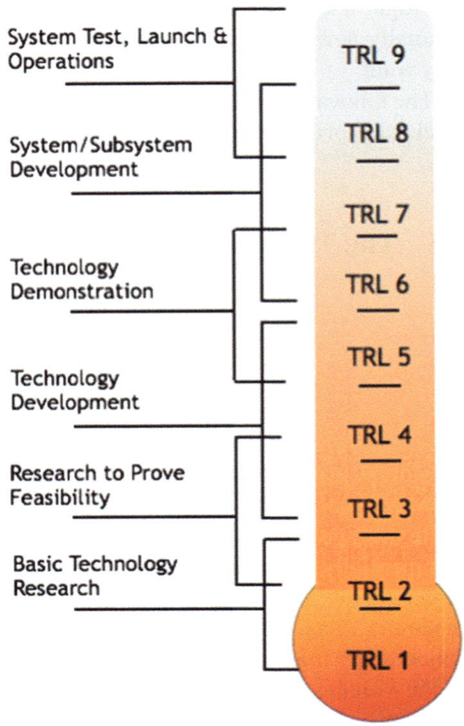

Fig. 4.2 NASA/DoD technology readiness level

and to judge if it can be acquired and implemented. It is a development of a complex technological system which needs the synchronized technological development of the all the components of its subsystem.

According to the NASA Bulletin Oct. 28, 2012, last updated in July 31, 2015 [24] *A technology project is evaluated against the parameters for each technology level and is then assigned a TRL rating based on the project progress. There are nine technology readiness levels, where TRL 1 is the lowest and TRL 9 is the highest.*

The synthetic definition and meaning of every step are given in the cited document:

- **TRL 1 Basic principles observed and reported**: Transition from scientific research to applied research. Essential characteristics and behaviors of systems and architectures. Descriptive tools are mathematical formulations or algorithms.
- **TRL 2 Technology concept and/or application formulated**: Applied research. Theory and scientific principles are focused on specific application area to define the concept. Characteristics of the application are described. Analytic tools are developed for simulation or analysis of the application.
- **TRL 3 Analytic and experimental critical function and/or characteristic proof-of-concept**: Proof of concept validation. Active Research and Development (R&D) is initiated with analytic and laboratory studies. Demonstration of

technical feasibility using breadboard or brass-board implementations are utilized with representative data.

- **TRL 4 Component/subsystem validation in laboratory environment**: Standalone prototyping implementation and test. Integration of technology elements. Experiments with full-scale problems or data sets.
- **TRL 5 System/subsystem/component validation in relevant environment**: Thorough testing of prototyping in representative environment. Basic technology elements integrated with reasonably realistic supporting elements. Prototyping implementations conforming to target environment and interfaces.
- **TRL 6 System/subsystem model or prototyping demonstration in a relevant end-to-end environment (ground or space)**: Prototyping implementations on full-scale realistic problems. Partially integrated with existing systems. Limited documentation available. Engineering feasibility fully demonstrated in actual system application.
- **TRL 7 System prototyping demonstration in an operational environment (ground or space)**: System is at or near scale of the operational system, with most functions available for demonstration and test. Well integrated with collateral and ancillary systems. Limited documentation available.
- **TRL 8 Actual system completed and *mission qualified* through test and demonstration in an operational environment (ground or space)**: End of system development. Fully integrated with operational hardware and software systems. Most user documentation, training documentation, and maintenance documentation completed. All functionality tested in simulated and operational scenarios. Verification and Validation (V&V) completed.
- **TRL 9 Actual system *mission proven* through successful mission operations (ground or space)**: Fully integrated with operational hardware/software systems. Actual system has been thoroughly demonstrated and tested in its operational environment. All documentation completed. Successful operational experience. Sustaining engineering support in place

The scale can be adapted to the specific research, development and innovation programs, other than the aerospace ones, by adjusting the definitions (or understanding) of the TRLs levels.

Significant adaptations were done by EARTO (European Association of Research and Technology Organizations), whose core mission is to harness science and technology in the service of innovation, to improve quality of life and build economic competitiveness. According to their mission, *ERTOs occupy nodal positions within innovation eco-systems, bringing together key players across the entire innovation chain, from fundamental to technological research, from product and process development to prototyping and demonstration, and on to full-scale implementation in the public and private sectors* [11].

Table 4.3 Scale used in Horizon 2020 work program of European union

Level	Description
TL 9	Actual system proven in operational environment
TL 8	System completed and qualified
TL 7	System prototype demonstration in an operational environment
TL 6	Technology demonstrated in relevant environment
TL 5	Technology validated in relevant environment
TL 4	Technological validity in a laboratory
TL 3	Experimental proof of concept
TL 2	Technology concept formulated
TL 1	Basic principle observed

The following Table 4.3 shows a reduction of the original NASA and DoD table to the table on the maturity of technology adopted in the Horizon 2020 work program of the European Union.[10]

Having the task to support and promote innovation, the European Union Commission has modified the original TRL scale, by taking into account not only the technological aspects, but also the market and organizational issues that can be crucial for innovation and for commercial success.

In a further simplified elaboration the scale is reduced to six stages. They represent the main points of the design process that have been analyzed in the book chapters.

The six phases of project management according to the previous table (Table 4.4) are:

1. Invention
2. Concept validation
3. Prototyping and incubation
4. Pilot production and demonstration
5. Market introduction
6. Market expansion.

Innovation needs the development of all phases. If one phase is missing, any innovation system collapses.

[10]Horizon 2020 is the biggest EU Research and Innovation program ever developed, with nearly 80 billion € of funding available over 7 years (2014–2020), in addition to the private investment that this money will possibly attract. It promises more breakthroughs, discoveries and world firsts by taking great ideas from the lab to the market. Horizon 2020 is the financial instrument implementing the Innovation Union, an Europe 2020 flagship initiative aimed at securing Europe's global competitiveness. Seen as a means to drive economic growth and create jobs, Horizon 2020 has the political backing of Europe's leaders and the Members of the European Parliament. They agreed that research is an investment in the future and so put it at the heart of the EU's blueprint for smart, sustainable and inclusive growth and jobs [8].

Table 4.4 A modified TRL scale incorporating manufacturability and non-technological aspects with a synthesis in six stages

Level	Action	Description
TL 9	Market expansion	Production and product fully operational
TL 8	Market introduction	Manufacturing fully tested, validated and qualified
TL 7	Pilot production demonstration	Low scale pilot production demonstrated
TL 6	” ” ”	Pre-production product
TL 5	Prototyping and incubation	Testing prototype in user environment
TL 4	Concept validation	Validation integrated prototype in lab environment
TL 3	” ”	First assessment feasibility of concept and Technologies
TL 2	Invention	Technology concept formulated
TL 1	”	Basic Principle observed

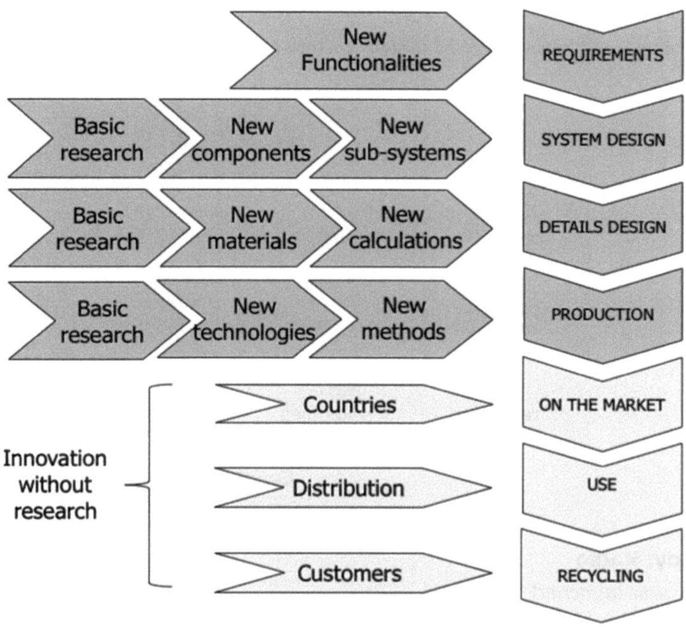

Fig. 4.3 Examples of innovation in a design process

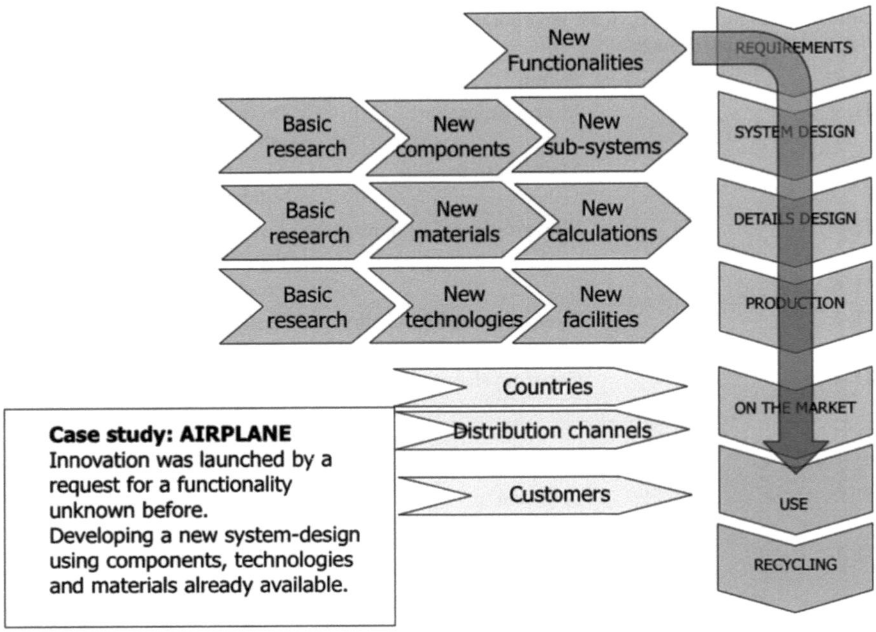

Fig. 4.4 Examples of innovation in the design of a new model of airplane

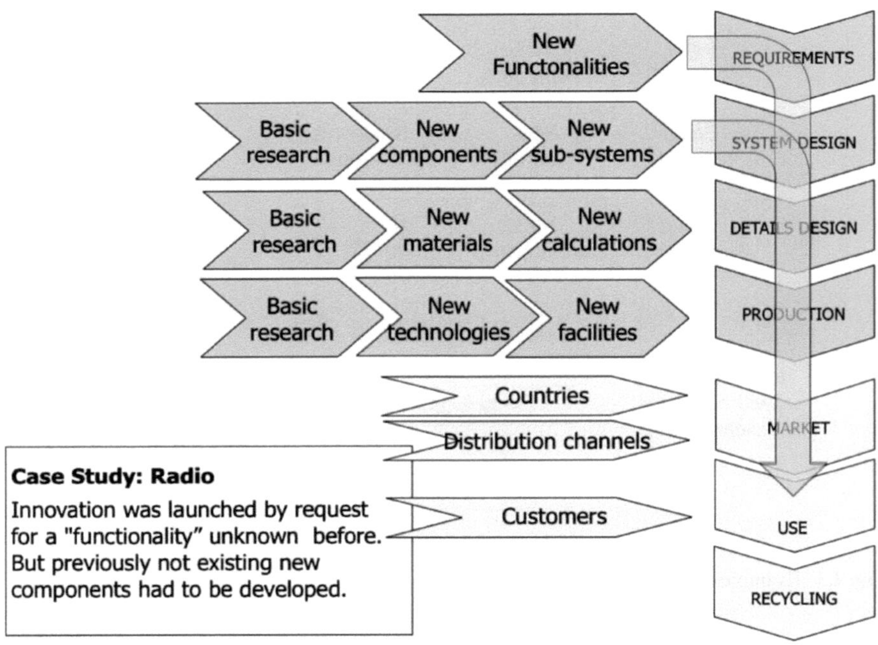

Fig. 4.5 Examples of innovation in the design of a new radio

4.6.1 Innovation Examples

The example of NASA activity in defining the TRL shows that innovation can be strongly influenced by public procurement (Figs. 4.3, 4.4 and 4.5).

Example 4.3 (*Innovation in the Design Process*)
Figure 4.6 shows the innovation scheme for a Morris Mini-Minor car, that in 1959 introduced innovative front-wheel drive and a transverse engine layout.[11]

Figure 4.7 shows another example of innovation introduced by the use of new sales channels and worldwide sales.

If a step of the design development is innovated, it results in a general innovation:

Functions Functionality not previously existing, i.e. ABS (to avoid locking wheel), etc.

System Design Innovating the design system, e.g. developing a new block-diagram: making a product with a new scheme (some industrial examples: independent suspensions, monocoque vehicle body, smart-phone, laser printer, etc.)

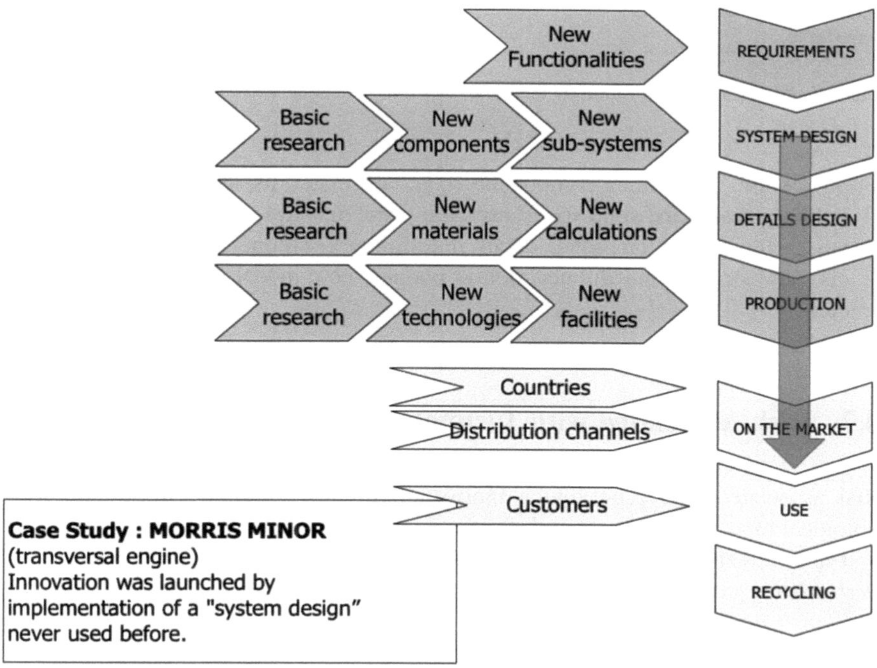

Fig. 4.6 Examples of innovation in the design of Morris Mini-Minor car

[11]Produced from 1959 until 2000 by BMC.

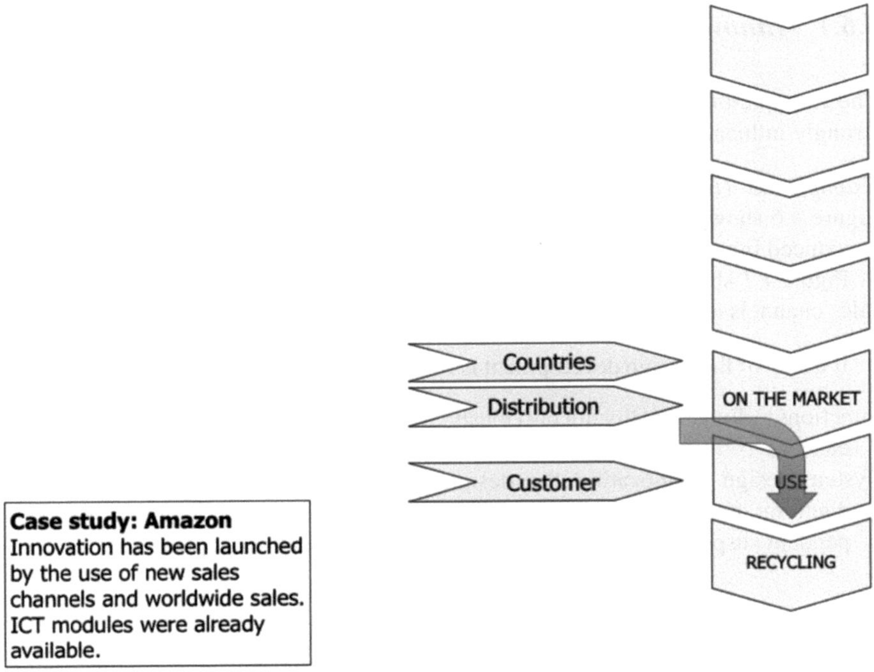

Fig. 4.7 Examples of innovation by the use of new sales channels and worldwide sales

Components Innovating components (e.g.: new materials, diesel engine introduction in private cars, solid state Laser, flat panel displays, etc.)

Details Innovating the project detail: redesigning components (e.g.: new car models, new smaller and thinner mobile phones, new model of washing machine, etc.).

4.7 Risk Associated with Technological Innovation

Risk associated with technological innovation implies that there will be some technological innovation that can fail [27].

Figure 4.8 shows the risk associated with the innovation in a *totally new, similar and existing* component, with respect to conditions of use that can be *the same, similar or new*, i.e. *totally different*.

Innovation may concern different levels of the project development: with the addition of features that did not exist before (level A), at system design (level B), at components design (level C) or at details design (level D). All these innovations can then be introduced at vary intensity degrees: low, medium and high.

Component	Use		
	Same conditions	Similar conditions	New conditions
New	Medium 2	High 3	Very high 3+
Similar	Low 1	Medium 2	High 3
Existing	0	Low 1	Medium 2

Fig. 4.8 Risk associated with technological innovation (from EFESO CONSULTING srl)

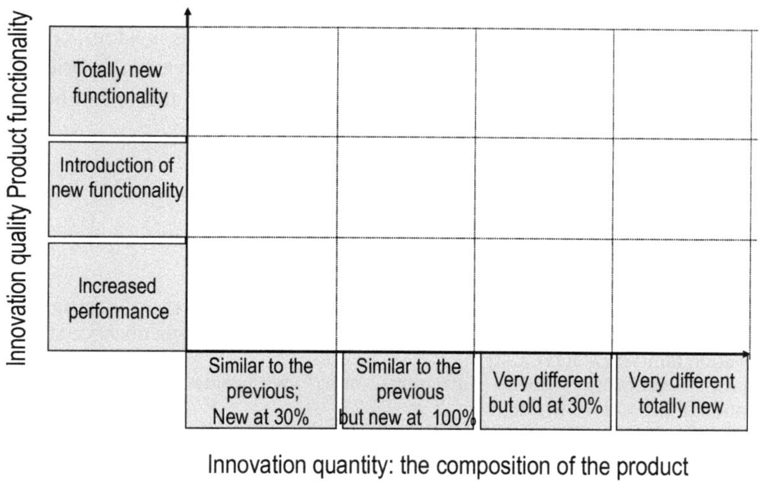

Fig. 4.9 Quality of innovation versus quantity of innovation (from EFESO CONSULTING srl)

Figure 4.9 shows a Table to assess the relationship between the quality of the innovation and the quantity of innovation introduced in a project.

4.8 What We Have Learned from This Chapter

We have seen the difference between a new scientific or technical idea (invention) and the process of translating an idea into a good or service that creates value (innovation). It is also possible to innovate something that already is on the market, providing elements of innovations in existing product and/or in existing production process and/or in service.

The pervasive dimension of innovation is proved by the concept of nation's competitiveness, when all the productive forces of a nation, in dependence on the capacity of its community, are oriented to innovation and upgrade. The responsibility of innovation is not only of the personnel directly involved but also of the cultural and political environment.

To innovate, strategies are needed. They have been defined over the years and provide a broad mass of measures and actions that are shown through examples.

In a time of commercialization of research results, TLR is a necessary method to allow a company to evaluate the readiness level of a certain technology, and to judge if it can be successfully acquired and implemented. Significant adaptations from the original version of these parameters were done by the European Association of Research and Technology Organizations, and so a modified TRL scale was proposed. The aim is to offer a tool to harness science and technology in the service of innovation. Several ways are shown to improve the quality of life and to build economic competitiveness, through examples of innovation of functions, system, components and details. Finally, risk associated with technological innovation is briefly evaluated, with the obvious implication that some innovations can fail or not be accepted by the market.

References

1. Branic B (2005) From time-to-market to time-to-volume. Portable Des 11(11):32
2. Business Dictionary (2016) Innovation. http://www.businessdictionary.com/definition/innovation.html#ixzz4JOVOCsfj
3. Business Dictionary (2016) Invention. http://www.businessdictionary.com/definition/invention.html#ixzz4JUwhJd6m
4. Business Dictionary (2016) Off-the-shelf-item. http://www.businessdictionary.com/definition/off-the-shelf-item.html#ixzz4JrmZfRqE
5. Business Dictionary (2016) Off-the-shelf-item. http://www.businessdictionary.com/definition/commercial-off-the-shelf-COTS.html#ixzz4JroABNGe
6. Christensen CM, Raynor ME (2003) The innovators solution : creating and sustaining successful growth. Harvard Business School Press. ISBN: 1578518520
7. Cohen MA, Eliasberg J, Ho TH (1996) New product development: the performance and time-to-market tradeoff. Manag Sci (1):173–186. https://doi.org/10.1287/mnsc.42.2.173
8. Commission E (2014) The EU framework programme for research and innovation. https://ec.europa.eu/programmes/horizon2020/en/what-horizon-2020
9. Consulting R (2006) About BOT model. http://www.ronghai.com/ronghai.jsp?pageid=ebot
10. Crawford MC, Di Benedetto A (2002) New product management, 7th edn. McGraw Hill, Boston
11. EARTO (2016) European associations of research and technology organisations. http://www.earto.eu/about-rtos.html
12. Jarvis A, Palmes P (2016) ISO 9001:2015 : understand, implement, succeed! Boston : Prentice Hall, uL: Order in West Room (Not borrowable) (2016.9.1883)
13. Kadareja A (2013) Risks of incremental, differential, radical, and breakthrough innovation projects. http://www.innovationmanagement.se
14. Lidwell W, Holden K, Butler J (2003) Universal principles of design. Rockport Publication, 1
15. McDougal J, Young WE (1995) Shortening the time to volume production on high-performance standard cell asics, pp 91–96. Hewlett-Packard
16. OECD (2005) Oslo manual guidelines for collecting and interpreting innovation data. DOIurlhttps://doi.org/10.1787/9789264013100-en. http://www.oecd.org/sti/oslomanual (PDF)

17. Ohno (2009) Definition from techtarget. http://searchmanufacturingerp.techtarget.com/definition/lean-production
18. Porter ME (2008) The five competitive forces that shape strategy. Spec Issue HBS Centen Harv Bus Rev 86(1):78–93
19. Porter ME (2016) Harvard business school: faculty and research. http://www.hbs.edu/faculty/Pages/profile.aspx?facId=6532&facInfo=pub
20. Rosenau MD, Griffin A, Castellion GA, (Author) NFA (1996) The PDMA handbook of new product development, vol 1, 1st edn. Wiley (27-09-1996)
21. SearchERP (2017) Lean production. http://searcherp.techtarget.com/definition/lean-production
22. Spacey J (2017) Simplicable guide. https://simplicable.com/new/time-to-volume
23. TRL (2012) Definition of technology readiness levels. https://esto.nasa.gov/files/trl_definitions.pdf
24. TRL (2015) Technology readiness levels. https://www.nasa.gov/directorates/heo/scan/engineering/technology/txt_accordion1.html
25. Tushman M, O' Reilly C (1996) Ambidextrous organizations: managing evolutionary and revolutionary change. Calif Manag Rev 38(4):8–29
26. Urban GL, Weinberg BD, Hauser JR (1996) Premarket forecasting for really-new products. J Mark 60:47–60
27. Venuvinod PK (2011) Technology, innovation and entrepreneurship Part III: My startup. Kluwer Academic Publishers. ISBN-13: 978-1-105-11627-8

Part II
Special Topics

The following three chapters deal with specialized topics at a basic level, which, however, cannot ignore mathematical modelling, even in an introductory presentation like this.

Chapter 5 introduces some safety concepts and reliability theory tools (such as FMEA). Chapter 6 deals with the design of experiment. Chapter 7 focuses on some basic concepts of the Taguchi method for robust design.

Chapter 5
Safety Assessment

Abstract Planning is a process which leads from acquired knowledge to new knowledge. Unknown materials, models, environments etc. can generate abnormal operating conditions with the risk of damage. Designers have a number of tools to protect themselves from this eventuality. The first concept that was introduced to guarantee the safety of a project is the safety factor. In general terms, a system has a safety factor when the collapse is expected with conditions markedly more rigorous than those provided as design requirements. However, in a vision closer to reality, the most correct way of addressing the problem consists in recognizing the probabilistic nature of each process in use of a product, involving not only *artificial*, but also human actions. Every product/service is naturally subjected to failures, but more than failure in itself, are the effects of a failure that give significance to the failure. Then not only the nominal functions (with no failures) but also the functions consequently to a failure must be considered as possible system's operating conditions. In an initial presentation there will be set forth some basic concepts of system reliability as the probability of no failures, along with the tools most commonly used, such as FMEA which, by virtue of their semi-scientific content, take into account variables not necessarily "measurable" but, nevertheless, lead to a significant reduction of the risk associated with a "failure".

5.1 Design for Safety

A broad definition of factor of safety (valid for every system and artifact), is based on this truth: design requires dealing with unknowns derived by science knowledge limits and by uncertainties of data. Thus Factors of safety are used to offset the effects of unknown variables [13]. A system has a safety factor when the collapse is expected in working conditions markedly more rigorous than those provided as design requirements. This can be obtained:

- By adding or improving materials quality.
- By developing devices to ensure a fail safe life, even in case of a partial collapse of the system.

A. Freddi and M. Salmon, *Design Principles and Methodologies*, Springer Tracts in Mechanical Engineering, https://doi.org/10.1007/978-3-319-95342-7_5

- By adding components to the system to support the utilization that is more rigorous than that provided in the specification as a minimum value. This can be obtained, e. g. by designing the system according to a redundancy principle.
- By designing components on the base of very well experimented principles, able to tolerate partial damage.

Accepted that the factor of safety assumes a value greater than one, it is difficult to assign typical values for different systems and artifacts, inasmuch as they depend on the types of loading and materials. In fact its value is related directly to the ignorance of the behaviorial knowledge of the loading.

5.1.1 Design for an Infinite Life

It requires that the parameter representing the *state* of each component (such as, for example, in the case of structural strength, the highest equivalent stress), always remains below the material stress limit.

In this case, the component is expected to have an *engineering-infinite* life. This method applies for example to the sizing of mechanical parts statically or even cyclically loaded, provided the amplitude of stress is constant over time. Today, it is replaced in many cases (but not always) by more realistic design principles, since infinite life often leads to unacceptable over-dimensions. Nevertheless, it is still applied to components with high fatigue cycle numbers, such as fundamental parts of engines (combustion, electric, turbines), electronic long life components, structural parts of machines for long-term operation, and uniformly stressed transmission shafts.

5.1.2 Design for a Safe-Life

This principle requires that the system be built in such a way as to operate safely throughout its expected life. This requires:

- A clear indication of operating conditions, maximum expected life time and environmental factors.
- In-depth safety checks (e.g. by means of detailed numerical simulations).
- Controls during fabrication and assembly.
- Verifying the effects of possible anomalous events such as overloads.
- Detailed inspection plans.
- Experimental determination of the actual term of life until the first failure.

Safe-Life is then fixed as a fraction of total life, experimentally determined.

This principle is used e.g. for components subjected to variable load cycles with infrequent load peaks, such as:

- Car gears.
- Transmission shafts affected by periodic over-loads.
- Racing vehicles parts.
- Rolling or sliding bearings.
- Rotors of rotating machinery.
- Pressure vessels design.
- Some aeronautical components such as wings, etc.

To introduce more advanced design principles which give the best safety guarantees, it is appropriate to consider carefully the concept of fault tolerance and distinguish mechanical systems from electronic and computer systems, systems [3]:

> In application with high reliability, availability or safety requirements, items must be designed to be fault tolerant at components level and/or fault tolerant re-configurable at equipment and system level. This means that at system level, the item considered should be able to *recognize a fault* (failure or defect) and quickly reconfigure itself in such a way as to remain safe and possibly continue to operate with minimal performance loss (fail-safe, graceful degradation).

In the case of mechanical components it is not always possible to have a minimal performance loss, but it is important that the system remains in a safe conditions and it will be able to warn the operator of the presence of the damage.

There are two situations that differentiate on a practical level: *fail-safe and damage-tolerant design.*

5.1.3 Fail-Safe Design

The principle requires that the system be built in a way to operate safely despite partial malfunctions. This assumes the following conditions:

- In case of failure, the system maintains some functionality.
- After the failure, it provides safety for a certain period of time.
- The system gives information on the partial failure (through vibrations, leak loss, pressure drop, power drop, slow motion, sound emissions, electrical or visual signals).
- The influence of the damage caused by partial failure on other subsystems, and hence on global security must be excluded, by a proper theoretical or experimental estimation.

Examples:

- Aeronautical structures with devices or constructive solutions capable of interrupting sub-critical propagation processes of defects before reaching unstable propagation.

- Elastic joints that reduce their functionality before breaking, but continue to maintain a partial functionality.
- Devices capable of containing fragments of broken non-vital components.

This method, introduced in aeronautical engineering, spread to other areas. However, it does not provide simulations through theoretical damage models.

Example 5.1 (*Fail-safe solution with crack stopper for airliner*)

A paradigmatic example is given by the case of stress around the window corners that was found to be much higher than expected, "probably over 40,000 psi [276 MPa]," and stresses on the skin were generally more than previously expected or tested. This was due to stress concentration, a consequence of the window's square shape. The problem was exacerbated by the punch rivet construction technique employed. The windows had been engineered to be glued and riveted, but had been punch riveted only. Unlike drill riveting, the imperfect nature of the hole created by punch riveting may cause the start of fatigue cracks around the rivet. The principal investigator concluded, "In the light of known properties of the aluminum alloy D.T.D. 546 or 746 of which the skin was made and in accordance with the advice I received from my Assessors, I accept the conclusion of RAE that this is a sufficient explanation of the failure of the cabin skin of "Yoke Uncle" by fatigue after a small number, namely, 3,060 cycles of pressurization [20].

See other examples in [21].

5.1.4 Damage-Tolerant Design

It is possible to operate a damaged system if its safety is guaranteed for a fixed period of time (between two maintenance intervals) through appropriate estimations by theoretical and experimental models. A case is given by mechanical systems containing defects. A system can be considered safe if a prevision of life is carried out with adequate tools, and periodic inspections are also used to ensure that the damage remains within acceptable limits. In some cases we can consider this approach as a method of life-extension.

This procedure is adopted e.g. in aeronautical design, in the design of pressure vessels and rotors for turbo-groups, with regard to special cases of over-stressing. Life expectancy is based on the laws of propagation of internal defects, detected by ultrasonic analyzers. In the periodic maintenance phases, the faults are detected and accepted if the critical dimension of the defects is theoretically known and therefore the residual life of the rotor estimated. It is based on the clearly calculated fatigue cycles (corresponding to the starting and stops of the machine). After periodic inspections the system is placed back into operation.

5.1.5 Design with Redundancy

The principle of redundancy offers a further means of increasing safety by assuming the support of other systems. It is accomplished when some functions are duplicated or multiplied with units or components in parallel, after checking that the failure of a unit is not in itself dangerous and does not lead to damage of the other vital parts. Nevertheless, it is important to observe, that, from Birolini book [3]:

> Redundancy does not just imply a duplication of hardware, since it can be implemented at the software level or as a time redundancy. However, to avoid common cause and single-point failures, redundant elements should be realized (designed and manufactured) independently from each other [...]. Investigation on all possible failure (fault) modes during the design of fault tolerant equipment and systems is mandatory. This is generally done using failure modes and effects analysis (FMEA/FMECA), fault tree analysis (FTA), causes to effects diagrams or similar tools, supported by appropriate investigation models. [...] failure modes analysis is essential where redundancy appears, among other to identify the parts which are in series to the ideal redundancy (in the reliability block diagram), to discover interactions between elements of the given item, and to find appropriate measures to avoid failure propagation (secondary failures).[...] These and similar considerations apply in particular for equipment and systems with high reliability and/or safety requirements, as used e.g. in aerospace, automotive and nuclear applications, ...

The assumption of functions from other components or systems can be total or partial. Examples:

- More engines on the same airplane.
- Dual (or multiple) cables for lifting equipment and cable-way.
- Double power supply lines.
- Power supply with buffer batteries.
- By-pass circuits on hydraulic systems.
- Multiple telephone and telecommunication lines.

5.2 Deterministic Approach: Safety Factor

Design approach is based on the factor of safety.

In a structural case an Allowable Stress Design (or Working Stress Design) S is defined and the factor of safety is defined as

$$n = \frac{R}{S}$$

where: R = strength and S = stress. Both are deterministic values. For safe design, $R > S$, and therefore,

$$n > 1$$

Uncertain parameters are modelled by one characteristic value. S and R are nominal values not defined in statistical terms. Differences in uncertainty between basic parameters are not considered.

5.3 Probabilistic Approach: Safety Index

On the contrary, considering stress and strength stochastic variables, the safety evaluation must be performed from a different point of view: uncertain parameters are modelled by the mean values and the standard deviations between these variables, assuming normal distributions. For the Central limit Theorem the probabilistic and statistical methods that work for normal distributions can be applicable to many problems involving other types of distributions. Then the stochastic variables are assumed to be normally distributed.

The first step is collecting statistical data on geometry, loading and material properties in terms of probabilistic density function for stress f_S and for strength f_R. The cumulative distribution function of a real-valued random variable X is the function given by.

$$F(s) = P(< \hat{s})$$

where the right-hand side represents the probability that the random variable s takes on a value less than or equal to $\hat{s}$. The probability that s lies in the interval (a, b), where $a < b$, is therefore

$$P(a < s < b) = F(b) - F(a)$$

Between probabilistic density function f_s and the (cumulative) probability function F the following relationship exists, Fig. 5.1:

$$F = \int_{-\infty}^{s} f(x)dx$$

Then function f is equal to the derivative of F:

$$f(s) = \frac{dF(s)}{ds}$$

Geometrically, $F(s)$ is represented by the underlying area of the $F(s)$ curve, for each s value, s is a random variable that assumes the meaning of stress or strength. If we

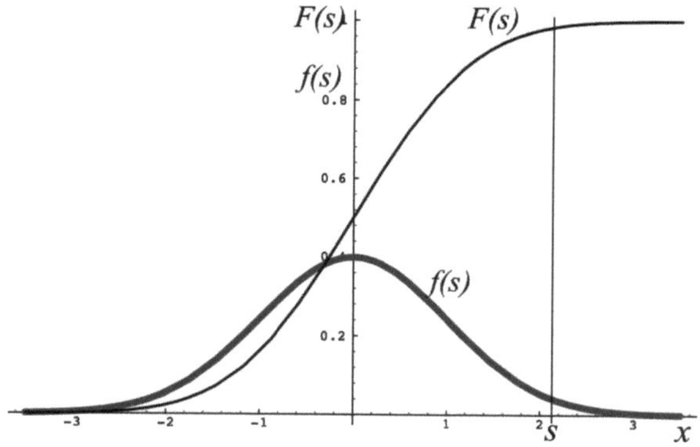

Fig. 5.1 Example of a probability density function $f(s)$ and the relative cumulative probability function $F(s)$

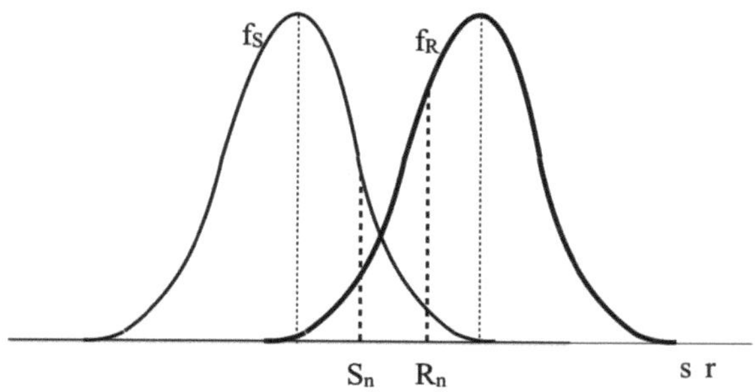

Fig. 5.2 Elementary verification by means of the safety factor: $S_n \leq \frac{R_n}{n}$

look for a link between the probabilistic and the deterministic setting, we observe that: a basic design requirement is given by the concept of Allowable Stress:

$$S_n \leq \frac{R_n}{n}$$

where S_n is the nominal stress and R_n is the nominal strength. If we limit the presentation to the most simple case of normal distribution, S_n would be larger than the mean of the probabilistic density function f_S of the stress (i.e. on the right side of the stress distribution) and R_n would be less than the mean of the probabilistic density function f_R of the strength (i.e. on the left side of the distribution), Fig. 5.2. To avoid failure the two nominal values must be at convenient distance.

The reason for a probabilistic interpretation is due to the uncertainties in the knowledge of stress and strength. For the stress state there are approximations in:

1. Stress analysis.
2. Peak loads evaluation.
3. Stress concentration evaluation.

As for strength, the uncertainties are related to:

1. Data on materials strength.
2. Size effects.

3. Manufacturing and, in general, production methods.
4. Assembly operations.
5. Aging effects.

The evaluation of failure probability for a simple system is performed by means of theoretical and experimental methods. For a simple structure, with a random load and with a random strength, some load-strength interference probabilistic models are utilized for covering:

- Static failure
- Fracture mechanics
- High cycle Fatigue
- Low Cycle Fatigue

Static Safety Index

The simplest approach to the load-strength interference probabilistic model is the *static limit state*, according to which failure takes place when a value of stress S exceeds a value of strength R, Fig. 5.2. The intersection area underlying by the two curves is the probability for which $S \geq R$.

For two stochastic *independent* variables (S and R) the *limit state function (or failure function)*, is defined as:

$$g = R - S$$

So the probability of failure is:

$$p_f = \textit{Probability that } [g \leq 0]$$

and the probability of failure, i.e. the intersection area is:

$$p_f = \int_0^{\infty} \int_0^{s} f_S(s) f_R(r) dr ds = \int_0^{\infty} f_S(s) F_R(s)\, ds$$

where $f_S(s)$ and $f_R(r)$ are the probability density functions of the two random variables R = strength and S = stress and $F_R(s)$ is the cumulative probability.

The *limit state function* may be defined in different ways, according to the assumption on the types of probability density functions.

In the following we will show only the calculation of the failure probability in the special but widespread case of both normal distributions. The calculation is simplified because a linear combination of normal functions is still normal and the intersection of the two curves is reduced to the intersection of a single standard normal curve with a straight line.

Safety Evaluation with Normal Distributions

When the limit state function is a linear combination of **normal** distributions, it can be represented as:

$$g = A + \sum_{i=1}^{k} B_i X_i$$

$$\mu_g = A + \sum_{i=1}^{k} B_i \mu_i \qquad \sigma_g = \sqrt{\sum_{i=1}^{k} B_i^2 \sigma_i^2}$$

where A and B_i are constants.

For the special case of two parameters S and R we have the following expressions for the mean and for the standard deviation of the function g:

$$\mu_g = \mu_R - \mu_S \qquad \sigma_g^2 = \sigma_R^2 + \sigma_S^2$$

We can reduce the two distributions to a unique normal standard distribution and calculate the failure probability with respect to the zero value.

$$p_f = P(g \leq 0) = P\left(\frac{g - \mu_g}{\sigma_g} \leq -\frac{\mu_g}{\sigma_g}\right)$$

Then the probability of failure assume this form:

$$p_f = \Phi(-\beta)$$

where Φ is the cumulative probability *cdf* of normal standard, with $\beta = \dfrac{\mu_g}{\sigma_g}$.

Now β is the parameter that synthesizes the safety estimation and is called *safety index*. Figure 5.3. The failure probability is represented by the area shown in Fig. 5.3 on the left side of zero. For $\beta = 0$ the area becomes half of the total area under the normal curve, and failure probability is 0.5.

In general, the choice of the distribution type is connected with the experimental data distribution. Consequently, the limit state can be different, and so are the reliability previsions.

Example 5.2 (*Probabilistic calculation of a beam*)
Let us assume the following values, Fig. 5.3 [8] R = N(62, 6.2) Q = N(100, 20) A = 2.8[1] Then:

$$\mu_g = \mu_R - \mu_Q/A = 26.286$$

[1]In practical applications we must consider only the case of values of stress and strength far from the respective mean values.

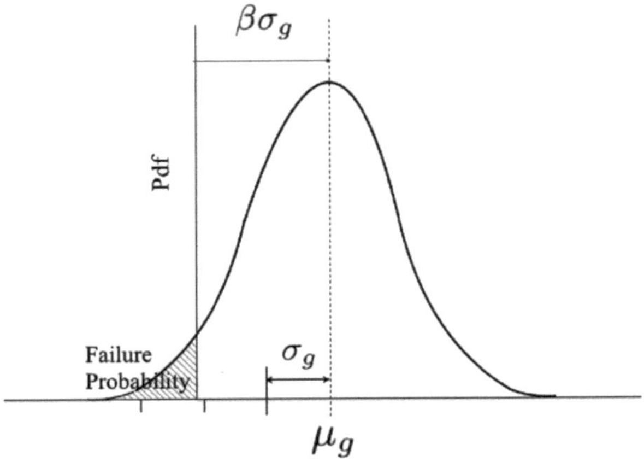

β	Failure Probability
0	0.5
1	0.158655
2	0.02275
3	0.00135
4	3.17E-05
5	2.87E-07

Fig. 5.3 Failure probability for different safety index values β

$$\sigma_g = \sqrt{\sigma_R^2 + \sigma_Q^2 \left(\frac{1}{A}\right)^2} = 9.458$$

Reference to a standard normal distribution equivalent for calculating the probability of failure:

$$p_f = P(g \leq 0) = P\left(\frac{g - \mu_g}{\sigma_g} \leq -\frac{\mu_g}{\sigma_g}\right)$$

therefore:

$$p_f = \Phi(\beta)$$

where Φ is the CDF and the safety index is $\beta = \mu_g/\sigma_g$

$$\beta = 26.286/9.458 = 2.78$$

This value corresponds to a probability of failure $p_f = 0.0027$

For illustration for the general case with random variables and analysis methods in reliability analysis and also for Monte Carlo Simulation see [22] with related bibliography, an application example in [11] and to other books.

In summary:

The safety index approach is preferable because:

- Statistical data can be included into design calculations.
- Better comparisons can be made between competing designs.
- Better refinement of the probability calculation due to improvement of the confidence in previous designs is a further advantage. Safety indexes are more effective than safety factors, that are of much more uncertain and empirical evaluation.

5.4 Failure Rate, MTTF and MTBF

We must point out the following.

For structures and structural components the classical reliability analysis is of limited use because:

- In the cases of structural safety analysis, reliability is assessed as the probability of failures that normally take place when stress exceeds the strength. Consequently, the probabilistic models are established for stresses S (derived from loads) and resistances R, both modeled as a function of time.
- Vice versa, in systems with *a large number of (semi-) identical components* subject to the same exposure conditions, the probability of a component failure is due to a steady deterioration as a function of time/use and must be assessed in a *frequentistic* manner from observed failure rates. In this case, the main point is *the statistical modeling of time until failure* [7, 12]. See also Table 5.1.

For the following subject, we will refer again to the presentation in the book [3].

If n are new and identical independent items put in operation at time $t = 0$ under the same conditions, and at the time t a number of these units $\overline{\nu}(t)$ have not yet failed, we can represent $\overline{\nu}(t)$ as a right continuous decreasing step function of variable t, Fig. 5.4.

$t_1, t_2, \ldots, t_n$ are observed failure-free times (i.e. operating times to failure) of the n items considered. The variable $\widehat{E}[\tau]$ given by the following equation, is the empirical mean, i.e. a statistical estimate, of τ:

$$\widehat{E}[\tau] = \frac{t_1 + t_2 + \cdots + t_n}{n} \tag{5.1}$$

The true mean $E[\tau]$ is obtained by the limit of this function for $n \to \infty$ and takes the obvious name of $MTTF$ = mean time to failure.

Given a n number of total items at time t and $\overline{\nu}(t)$ number of not filed items at time t the empirical reliability is given by:

$$\widehat{R}(t) = \frac{\overline{\nu}(t)}{n} \tag{5.2}$$

For $n \to \infty$ it holds that:

$$\lim_{n\to\infty} \widehat{R}(t) = R(t) \tag{5.3}$$

where $R(t)$ is the true reliability.

For a time interval $t, t + \delta t$, the empirical failure rate is defined as:

$$\hat{\lambda}(t) = \frac{\overline{\nu}(t) - \overline{\nu}(t + \delta t)}{\overline{\nu}(t)\delta t}$$

In this equation $\hat{\lambda}(t)$ is the ratio of the items failed in the interval $t, t + \delta t$ to the number of items still surviving at the time t. Dividing numerator and denominator by n gives:

$$\hat{\lambda}(t) = \frac{\widehat{R}(t) - \widehat{R}(t + \delta t)}{\widehat{R}(t)\delta t}$$

if R is derivable, for $n \to \infty$ and $\delta t \to 0$ $\hat{\lambda}(t)$ converges to the instantaneous failure rate $\lambda(t)$

Table 5.1 Failure rate values for 10^6 operating life (Hours). On the left side (Atomic Energy of Canada Ltd.). Modified from [10]

Machine components	min	max
Couplings		
Elastomers	20	30
Disks/diaphragms	0.01	0.1
Gears		
General use	8	50
Helicoidal	0.5	15
Shafts		
Low loads	0.02	0.1
High loads	0.1	0.5
Crankshafts	5	8
Clutches	2	10
V- Belts	20	80
Springs		
Low loads	0.01	0.1
High loads	0.8	2.5

Bearings	min	max
Bushing	4	10
Spheres	5	50
Roller bearings	3	10
Seals		
O ring	0.1	0.7
Elastomeric	8	10
Metallic	25	200
Valves		
In pressure vessels	50	150
Manual valves	0.4	6
Fastening elements		
Spines	0.001	0.007
Rivets	0.001	0.01
In Cylindrical bodies	0.01	0.1
Of pump housings	0.01	1.0

Anti-vibration elements	min	max
elastomeric	6	20
Electric motor windings	8	20
Small motors	10	20
Great motors	5	10
Vehicles components	[5]	
Brakes	13	
Clatches	2	
Compressors	65	
Differential gear	15	
Fans	6	
Heat exchangers	4	
Gears	0.2	

Pomp	12
Springs	5
Valves	14
Tools	26
Battery	0.6
Alternator	2
Generator	40
Heating circuit	4
Lamps	
Incandescent	10
Neon	0.5
Solenoid	1
Switch	6

The failure rates refer to standard test conditions and a standard quality level. These base rates have to be adjusted with appropriate coefficients when they refer to different quality and use conditions:

$$\lambda_{Eff} = \lambda \cdot \pi_Q \cdot \pi_A \cdot \pi_K$$

where:

λ_{Eff} = effective value

λ = base value rate

π_Q = quality factor

π_A = environment factor

π_K = use factor

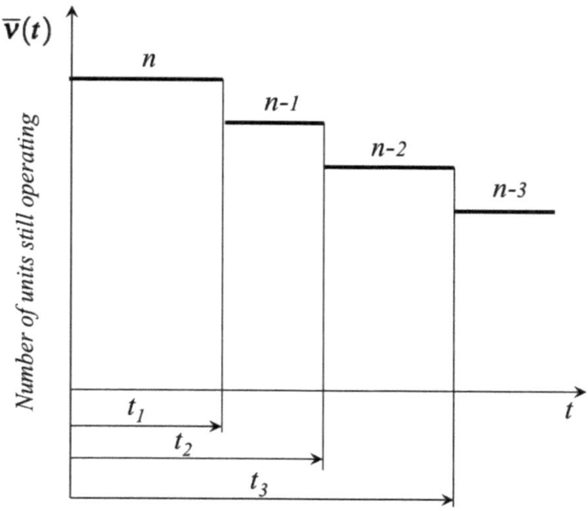

Fig. 5.4 Number of items still operating at time t

$$\lambda(t) = -\frac{dR(t)}{R(t)dt} \tag{5.4}$$

Rearranging the Eq. 5.4 and integrating for x from 0 to t we obtain:

$$R(t) = e^{-\int_0^t \lambda(x)dx} \tag{5.5}$$

considering $R(0) = 1$ i.e. for $t = 0$ all items are new.

In many practical application $\lambda(x)$ is constant and R assumes the form of a negative exponential distribution:

$$R(t) = e^{-\lambda t} \tag{5.6}$$

Reliability can be defined as the probability, at a given confidence level, that the equipment will perform their intended functions without failure and with specified performance limits, at a given age, for a specified length of time (the mission time), when used in the manner and for the purpose intended, while operating under a specified stress level. Inversely the failure probability in the same time interval is:

$$F(t) = 1 - e^{-\lambda t} \tag{5.7}$$

Probability density function is given by:

$$f(t) = \frac{dF(t)}{dt} = -\frac{dR(t)}{dt} = \lambda(t)R(t)$$

Furthermore, the mean (expected value) of the failure-free time $\tau > 0$, defined as mean time to failure MTTF, is given by.[2]

[2]The demonstration in the following steps:

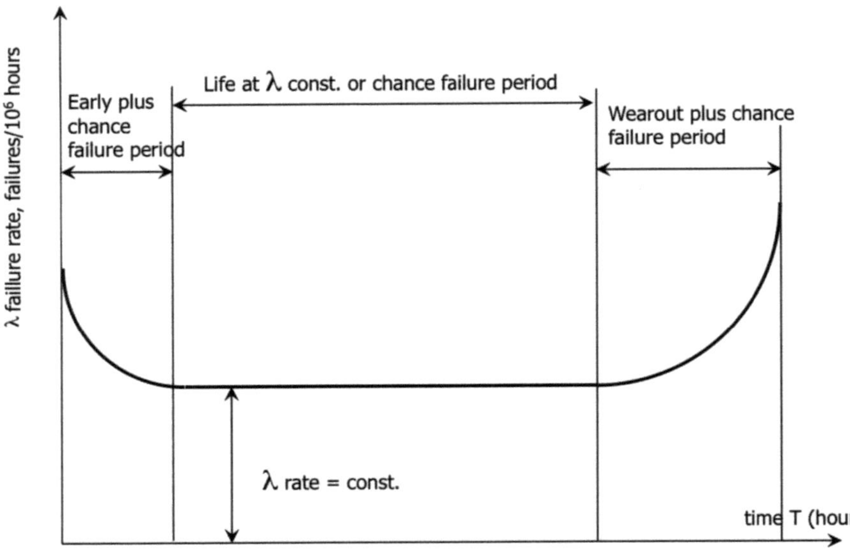

Fig. 5.5 Operating life (Hours)

$$MTTF = E[\tau] = \int_0^\infty tf(t)dt = \int_0^\infty [1 - F(t)]dt = \int_0^\infty R(t)dt \tag{5.9}$$

For the Eq. 5.6:

$$MTTF = 1/\lambda \tag{5.10}$$

In this case the failure rate λ can be estimated by the relationship:

$$\hat{\lambda} = k/T \tag{5.11}$$

where T is a given cumulative operating time and k is the total number of failures during T. Also for repairable items with a constant failure rate, if the item, after a each repair is considered as good as new, the successive failures-free times are considered independent random variables like for a new item. Then, only in this case:

$$MTBF = 1/\lambda \tag{5.12}$$

where $MTBF$ stands for mean operating time between failures.

The failure rate of a large population of identical items has a typical representation in a bathtub curve, Fig. 5.5.

$$MTTF = -\int_0^\infty t\frac{dR}{dt}dt = -[tR]_0^\infty + \int_0^\infty R(t)dt = \int_0^\infty R(t)dt \tag{5.8}$$

The previous demonstration is valid only for $E[\tau] < \infty$. See a discussion on this point in Example A 6.9 of [3].

The early failures of the first part are due to randomly distributed weaknesses of materials, components or production processes. The central part of the curve describes failures distributed according to Poisson model and in the third part failures are due to wear out i.e. aging, fatigue, corrosion etc.

- In practice, the mean time between failures ($MTBF = 1/\lambda$) is often reported instead of the failure rate. This is valid and useful if the failure rate may be assumed constant (often used for complex units/systems and for electronics) and is a general agreement in some reliability standards (Military and Aerospace).
- *MTTF* or *MTBF* are only comparison parameters. See Table 5.1 for several rate values of machine components. We must not forget, that, with the constant rate hypothesis, they relate only to the flat region of the bathtub curve, also called the "useful life period". Because of this, it is incorrect to extrapolate *MTBF* to give an estimate of the total service life time of a component, which will typically be much shorter than suggested by the *MTBF*, due to the much higher failure rates in the *end-of-life wear-out* part of the *bathtub curve*.
- Early failures are not defects or systematic failures caused deterministically by errors or mistakes. These should be removed by a full change in design and/or in production process.
- The *MTBF* appears frequently in the engineering design requirements, and governs frequency of required system maintenance and inspections.

5.5 Reliability of Systems with Simple Structure

Simple Structures have a reliability representation (reliability block diagram RBD) that can be reduced to a series or parallel form with independent elements. A RBD is drawn as a series of blocks. Each block represents a component of the system with a failure rate. The elements which are necessary for the required function are connected in series and if one of these fails, then the whole system fails. By contrast, all elements in parallel represent a redundancy, since all the parallel paths must fail for the parallel network to fail. A reliability block diagram basically differs from a functional block diagram, since one or more elements can appear more than once in a reliability block diagram, something that cannot happen in a functional diagram [3]. In the same source it is clear that each element represents only two states (good or failed) and *one failure mode*. Thus for the elements of a reliability block diagram:

- *In a series path* all the elements must not fail (survive) for the fulfillment of the required function. The system reliability is then the product of the reliability of each element:

$$R_s(t) = R_1(t) \cdot R_2(t) \cdot \ldots R_n(t) = \Pi_{i=0}^{n} R_i(t) \tag{5.13}$$

- *In a parallel path* all the elements must fail for the function failure. The failure probability for the system is then the product of the failure probability of each element:

$$P_s(t) = P_1(t) \cdot P_2(t) \cdot \ldots P_n(t) = \Pi_{i=0}^{n} P_i(t) \tag{5.14}$$

For the variety of reliability block diagram and the relative reliability functions, in the hypothesis of independent no-repairable elements up to system failure, new for $t = 0$, see Table 2.1 in [3].

Recalling the expression $R(t)$, from 5.13:

$$\lambda_S = \sum_{i=1}^{n} \lambda_i \tag{5.15}$$

Then the failure rate for a complex system is simply the sum of the individual failure rates of its components. This permits testing of individual components or subsystems, whose failure rates are then added to obtain the total system failure rate. All the systems that can be interpreted by means of a block diagram (RBD) with independent elements in series and in parallel are considered simple systems. For them, the reliability can be calculated if the failure rate for each element is known. If the overall reliability is not satisfactory, one must search for the weak points of every component of the system, reduce the failure rate of some component, repeat the calculation procedure re-evaluating the overall reliability.

Each function of the system can have its own reliability diagram.

5.6 Failure Mode and Effect Analysis

From the design point of view, up to this stage, we have seen the design process of defining positively what a design must be developed to satisfy the customer. Now we must complement that point of view, describing methods intended to recognize and evaluate deficiencies, i.e. potential causes of failures of a product or of a process and their effects and to develop the strategies to eliminate them.

These methods are, in fact, reliability design methods that extend the specification data to abnormal conditions of use and take into consideration the realistic behavior of the system, sub-systems and components of the overall design.

This section describes the Failure Modes Effects Analysis (FMEA), often referred to as FMECA (Failure Mode and Criticality Effect Analysis). The first is more qualitative and the second is more quantitative; both were developed in the late 1950s to study problems of malfunctions of military systems and were objects of a large number of applications and specializations which led to some confusion of their definition [8]. Here we will describe FMEA in a form of [9, 14, 17]. We suggest specialized books and manuals for different types of FMEA [6].

- For Conceptual Design: [9].
- For a Process design [1].
- For historical information [7, 19].

1. What is important is that FMEA is a bottom-up, inductive analytic method which may be performed at either the functional (called also software) or physical principle (called hardware) level. The functional approach is used when items are not fully defined but the function failures can be identified while, vice-versa a hardware approach is used when can be physically and geometrically defined.
2. For the successful implementation of an FMEA program, it should be a *before-the-event* action, i.e. the FMEA must be done before a design or process failures have been detected, or actually have taken place in reality. It is a prediction method, i.e. the estimation of *potential failure modes and the relative effects*. The emphasis, more than on the failure mode itself (weakness, deficiency, critical behaviour), is *the severity of the effects* of a failure mode.
   It must be considered an interacting process, as a never ending quality tool. FMEAs completed as soon as possible, can ensure not only a sensitive reduction of failures but also easier and less costly changes and smooth initial production.
3. FMEA is not a pure scientific method, since it utilizes human experience and expertise, which has been historically accumulated from similar cases and based on the scientific/technical memory.
4. Abnormal operative conditions (and not only the nominal ones) can be considered as potential causes of failure modes, together with other variables depending on the context such as laws and human factors etc. (The treatment of stochastic variables as disturbances and their effects will be the object of a specific analysis in Chap. 6).
5. The method can be applied at the design phase (DFMEA) and at production process phase (PFMEA) to the following different items:
   - Systems (also in the conceptualization stage, with the name of Concept FMEA), evaluating also the interaction with human factors.[3]
   - Sub-systems.
   - Assemblies.
   - Sub-assemblies.
   - Components or piece-parts.

   But these tools can be successfully applied for any *material and immaterial process* in order to critically analyze their propensity to failures, malfunctions, etc. as:

[3] We emphasize that the word "Concept", used in Product development phase [4] as a clear, detailed description of the attributes and benefits of a new product that addresses the needs of the targeted customers.

- Reliability-Centered Maintenance.
- Assembly Procedure.
- Instruction Manual.
- Service.
- Business Process.
- etc.

Before starting the illustration, we recall the rationale of it.

A risk evaluation can be done on three assumptions:

- A failure mode is serious or not serious not in itself but on its effects. This effects analysis is performed on the basis of logic, by evaluating (on the historical base of similar design or on reasonable previsions) the consequences of a failure. So we have to judge the effects *severity*.
- Then it is important to establish when the failure mode occurs, or rather, the *occurrence of its causes*. A logical thinking must go back to the causes sequence, to discover the first causes while, again, the occurrence estimation can be based on the history of similar failure mode in similar existing systems or on the theoretical/experimental knowledge of the failure rates for new systems.
- Finally it is important to determine which controls can be put in place, to foresee or, better, to prevent the occurrence of the failure. This can be done on the base of the knowledge of the effectiveness of the controls, or their *detectability* by means of a similar type of controls. These estimates are used to build a risk parameter. Once the risk is evaluated, it is of course also essential, to implement proper corrective actions.

Coming back to the procedure, these are the steps:

1. The first point is to discover the failure modes starting from a question: What could possible go wrong with this system? Clearly this is a (mal)functional analysis.[4]
2. As we go down to lower levels of sub-systems, assemblies etc., it is natural to move from a functional analysis to physical analysis, until the component level, for which the question become: *What* could possibly go wrong with component?.[5]

Example 5.3 (*The malfunction of a component is due to faults*)
With reference to the overall item function: *Grip the specimen* analyzed in Chap. 8, Fig. 8.5, the right hinge sub-functions of the couple *Forks and Pins* are:

- Allow the force transmission.
- Allow assembly and disassembly of the pin from the fork and the hole of the specimen (Fig. 5.6).

[4]The functions of the item can be identified by a QFD analysis, that must also include anomalous operational conditions.

[5]Potential failures modes are due to faults in components.

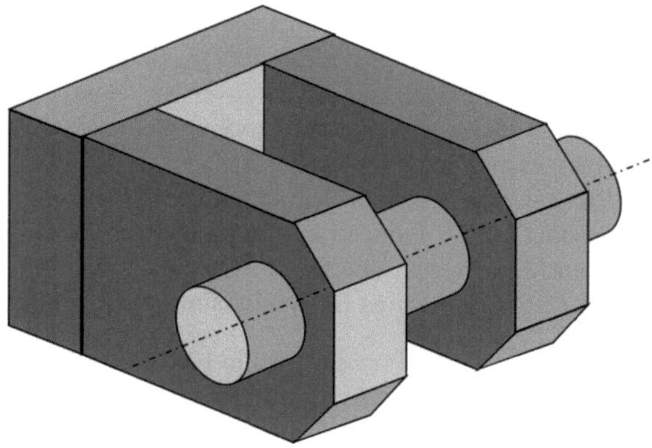

Fig. 5.6 The function of a correct assembly and disassembly of pin and holes is related to physical properties: stress, materials, geometrical and surface tolerances

The sub-functions failures can arise from the following *faults* of the components:

- Break of pin or fork.
- Global plastic deformation of the pin axis or the fork.
- Local plastic deformation of the contact area.

The *causes* of the faults could be:

- Excessive stress in pin and fork.
- Excessive Hertzian stress in the contact.
- Unsuitable materials of the parts.
- Unsuitable surface thermal treatments of the parts.
- Improper selection of geometrical tolerances.
- Improper selection of the surface roughness.

5.6.1 Who Does FMEAs

Since it is difficult for a designer to admit potential weaknesses on his/her work, FMEA should be developed by a team of personnel qualified to a constructively examine the design and process, avoiding frustration of the designer. A small team of a few (four) members is enough for developing an effective FMEA.[6]

A core Design FMEA Team can be formed by the following skills:

[6]This small team is a group of people with a full set of complementary skills required to complete a task, job, or project. Team members (1) operate with a high degree of interdependence, (2) share authority and responsibility for self-management, (3) are accountable for the collective performance, and (4) work toward a common goal. A team becomes more than just a collection of people when a strong sense of mutual commitment creates synergy, thus generating performance greater than the sum of the performance of its individual members [5].

1. *Product Engineer as Team Leader*. He/she must define problems without suggesting solutions, providing data on potential product failure modes, causes of failure occurrence and effects severity.
2. *Quality Engineer* with an historical memory of similar cases and expertise with controls and detection methods.
3. *Process Engineer* is the process expert. He/she must provide input on causes of product failure induced by the production process and improvement for manufacturing better products.
4. *Wild card Member* as facilitator, supervisor, with knowledge of similar products.

Vice-versa, the core Process Design FMEA is formed by: *Process Engineer as primary member and Leader*, who provides input on potential manufacturing failure modes and effects and the second member is a Quality engineer and the third is the *Product Engineer* that provides suggestions for avoiding to produce unacceptable products.

Coming back to product design, the team must answer if there are other ways than those considered by the designer, in which things can go wrong.

- How can this product (or process) fail?
- What effect will it have on the product or system if this failure occurs?
- What actions can prevent this failure?

From what has been learned, the clarification of all the potential failure modes of a design or of a process, and their relative causes, is a judgment based upon three parameters:

- Gravity of effects of a failure.
- Occurrence of the Causes of failure.
- Effectiveness of detection methods of failure modes and causes.

5.6.2 Design FMEA (DFMEA)

The first action to implement the method is the identification of all the failure modes.

- List of potential System/Product failure modes.
- List of the causes of system/product failure modes and the actions to eliminate (or reduce) them.
- List of System/Product functions to be monitored to detect potential failure modes.

5.6.2.1 Determining the Effects of Failure Mode and Their Severity

Each potential failure mode has its potential failure effect (*What* will happen if the failure mode occurs?). After the failure modes identification, our attention has now

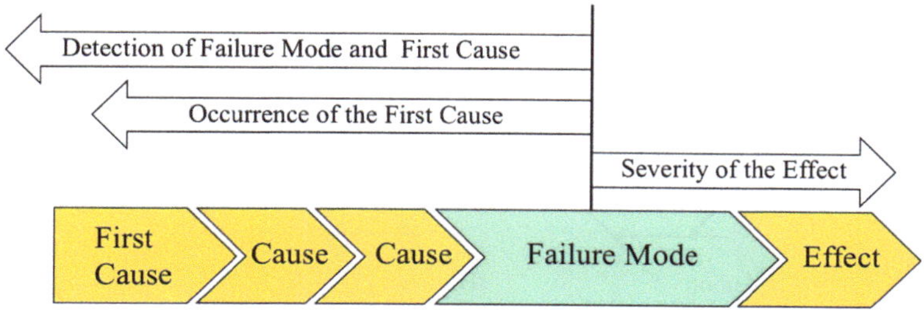

Fig. 5.7 The Risk analysis needs three perspectives

Table 5.2 Severity rating for systems/items failure mode effects [23]

Effect	Criteria	Ranking
None	No effect on customer	1
Minor effect on product	Slight customer annoyance	2 – 3
Significant effect on product	Moderate customers annoyance	4 – 5
Expensive repairs	High degree of customer annoyance	6 – 7 – 8
Potential effect on safety	Total loss of primary function	9[a]
Hazardous without warning	Catastrophic-safety related failure	10[a]

[a]Component or system receiving 9 or 10, appropriate corrective actions must be taken to reduce the rating.

turned downstream in the logical sequence of causes-failure modes- effects, Fig. 5.7 to the identification (and evaluation) of the effect.

If the item has more than one potential failure modes, all must be considered separately and each effect separately evaluated.

At this point the team must assess the seriousness (or severity) of each effect. The severity (S) must be estimated on a 1–10 scale, Table 5.2.

5.6.2.2 Identifying the Causes of Failure and Their Occurrence

If the first step is a failure mode, our attention has now turned back upstream, at the identification and evaluation of the potential cause, i.e. at the design weaknesses, the consequence of which is a failure mode.

The team should answer the question: *What conditions can cause this failure mode?* The team must assess the likelihood of occurrence (O) of each cause. The possibility that causes exist which could bring a potential failure must be estimated on a 1–10 scale.

Another problem arises from the fact that some causes are of first level, i.e. are the immediate causes (or major causes) of a failure mode, while others can be the causes of higher level, i.e. causes of the causes or root causes. One of the ways to show this dependence is a Cause and Effect Diagram. Typical structure is the Ishikawa

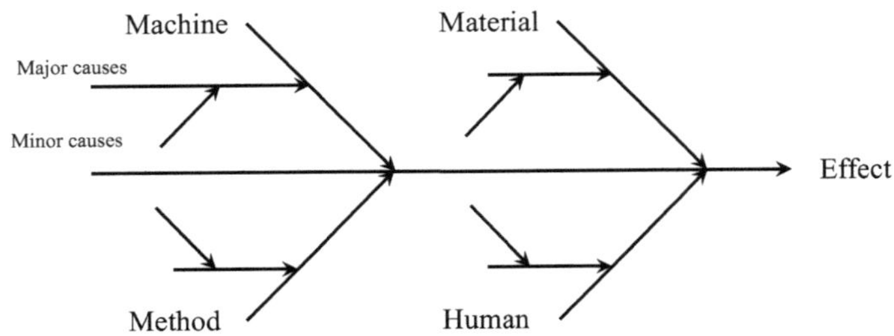

Fig. 5.8 Ishikawa or cause-and-effect diagram for the visualization of relationships between causes and effects (Modified from [3])

Table 5.3 Occurrence rating ([9])

Occurrence	Criteria	Failure rate	Ranking
Almost never	Failure unlikely	$\leq 1/1{,}500{,}000$	1
Remote	Rare number of failure likely	1/150,000	2
Very slight	Very few failures likely	1/15,000	3
Slight	Few failures likely	1/2,000	4
Low	Occasional no. of failure likely	1/400	5
Medium	Medium no. of failures likely	1/80	6
Moderately high	Moderately high no. of failures likely	1/20	7
High	High no. of failures likely	1/8	8
Very high	Very high no. of failures likely	1/3	9[a]
Almost certain	Failure almost certain to occur	$>1/2$	10[a]

[a]For component or system receiving 9 or 10, appropriate corrective actions must be taken to reduce the rating.

or Fishbone Diagram [3] where causes are grouped into Machine, Material, Method and Human (Man), into failure mechanisms, or into a combination of all of them, Fig. 5.8. So a reasonable doubt in this phase is what is really a failure mode. Is it a failure cause or a failure mode? Rules do not exist for a strict evaluation. If a primary cause can be identified, we start from the occurrence evaluation of that one. All the causes must now be rated on their occurrence. Table 5.3 can help give this evaluation, based on historical data of similar products.

5.6.2.3 Establishing Current Controls and Their Detection Likelihood

A Design control is a method to detect causes of the first level (of a potential failure mode) or of failure modes themselves. The team should answer the question: *Given a defect, how will it be detected?*

It is not necessary to distinguish, from a methodological point of views, between the FMEA on the conceptual solution of the system (Concept) and on the individual

Table 5.4 Detection rating

Effect	Ranking
Extreme likely	1
Highly likely	2 – 3 – 4
Likely	5 – 6 – 7
Unlikely	8 – 9
Extreme unlikely	10

component. If physical controls (tests) are possible on early prototypes or simulacra, they are preferable. When this is not possible, simulation tests by computer must be used. The first level of control is on *prevention* of causes and effects of a failure mode or reduction of their occurrence, then on *detection of causes* and suggestion of corrective actions, and finally on the *detection of failure modes themselves.* As a general rule, controls on the system are made by tests on prototypes and by numerical simulation.

Examples of current design controls on items are design reviews, mathematical models, fail-safe assessment, laboratory and in-field testing, prototype tests, material and structures tests etc. used in similar designs.

A very general scheme on the likelihood of detection (D) with the relative ranking from 1 to 10 is given in Table 5.4 (1 means almost certain likelihood of detection and 10 almost impossible).

5.6.3 Process FMEA (PFMEA)

Failure modes are not derived solely from design deficiencies. They derive also from the manufacturing process.

Design DFMEAs and Process PFMEA have complementary tasks:

- Design DFMEAs identify failure modes and resulting effects of a product design: for systems it helps to select the optimum design alternative for the overall function, and interactions with other systems, while for sub-systems helps primarily to identify the failure modes and the causes that generally are due to faults of single-piece component. *How might this system (sub-system, assemblies...component) function fail?* They help to develop optimum product design.
- Process PFMEAs focus on potential failure modes of sub-systems, (assemblies sub-assemblies, single-piece components) caused by manufacturing or assembly process deficiencies. *How might this process (of manufacturing, assembly) fail?* It focuses on potential product failure mode caused by process deficiencies. It helps to develop optimum manufacturing and assembly processes and assists in the analysis of new manufacturing and/or assembly processes.
- While Design DFMEA aims for the optimum fitness for use of the product, Process PFMEA points to the optimal conformity of the product to its design. Carried on

in parallel, their aim to achieve an optimal product with an optimal process. Obviously, not all process deficiencies have influence on product but Process PFMEAs are interested in the latter, Fig. 5.9 from [2].

- From a theoretical point of view the Process PFMEA has a very similar development as the Design DFMEAs already seen. The fundamental questions are the same as before, but it is useful to show some differences in the table of occurrence rating and of detection rating, since for PFMEA, criteria on the manufacturing process must be introduced, as the capability of the production process and probability of defect reaching the customer [1]. The last one is the obvious conclusion of the process phase that has its natural consequence in shipping of the product.

For the occurrence rating, e.g. Table 5.5 and for detection rating, see Table 5.6

How to estimate the causes and failures occurrence for new items not already experimented in previous designs? For the functional level, engineering judgment may be required to assign the ratio of the failure mode. For the hardware level, a criticality analysis is necessary. The concept of *quantitative criticality rating* is introduced [18]. A quantitative assessment needs the knowledge of the mission phase duration t and the estimation of the failure rate λ (for assemblies, sub-assemblies and components). It can be obtained by technical literature [16] or directly by an experiment designed for determining the reliability of the new item. If both parameters λ and t are correctly estimated, Eq. 5.7 can help to build a model for the estimation of a *failure mode criticality number*:

$$C_m = \beta \alpha \lambda_p t \tag{5.16}$$

where:

- λ_p is a basic failure rate.
- α is a failure mode ratio (for one item $\sum_{\alpha} = 1$).
- β is the conditional probability of loss of function or mission. It is the conditional probability that the failure effect will result in the identified severity classification, given that the failure mode occurs. It represents the analyst's best judgment of the likelihood that the loss will occur.
- t is the mission duration.

$\lambda_p t$ allows the failure probability calculation: $1 - e^{-\alpha \lambda_p t}$

The modal criticality number is calculated for each failure mode of each item (C_{mi}), and the criticality number C_r for each item is calculated as sum of the criticality numbers C_{mi} for all failure modes that are contemporarily present in a item

A very common example is the mechanical design of gear teeth: fatigue stress is accompanied by a wear. There are therefore two failure modes that compete with each other, with two different effects, each of them defined by a criticality parameter C_{mi}. For a discussions on FMEA and FMECA see [3, 7, 15].

Table 5.5 Occurrence rating ([9])

Probability of occurrence	Criteria	Failure rate	Ranking
Remote	Capability of ±4 within specs	$p < 1/20{,}000$	1
Low	Capability of ±3.5 within specs	$p < 1/15{,}000$	2
	Capability of ±3.5 within specification	$p < 1/10{,}000$	3
Moderate	Occasional failures. Capability of 3.0 within specs	$p < 1/2{,}000$	4
	Occasional failures	$p < 1/1{,}000$	5
	Occasional failures	$p < 1/200$	6
High	Frequent failures. Capability of 2.5 or less within specs	$p < 1/100$	7
	Frequent failures	$p < 1/20$	8
Very high	Failure certain	$p < 1/10$	9[a]
	Failure certain	$p > 1/10$	10[a]

[a]For component or system receiving 9 or 10, appropriate corrective actions must be taken to reduce the rating.

Table 5.6 Detection rating ([1])

Likelihood	Criteria	Probability of defect reaching the customer	Ranking
Remote	Likelihood that product would ship	$p < 1/10{,}000$	1
Low	Likelihood that product would ship. 100% automatic inspection. 100% operator controlled inspection	$p < 1/5{,}000$	2
Low	Likelihood that product would ship. 100% automatic inspection. 100% operator controlled inspection	$p < 1/2{,}000$	3
Moderate	Likelihood that product would ship. 100% automatic inspection of variable characteristics. Sampling with statistical analysis	$p < 1/1{,}000$	4
Moderate	Likelihood that product would ship. 100% automatic inspection of variable characteristics. Sampling with statistical analysis	$p < 1/500$	5
Moderate	Likelihood that product would ship. 100% automatic inspection of variable characteristics. Sampling with statistical analysis	$p < 1/200$	6
Moderate	Likelihood that product would ship. 100% automatic inspection of variable characteristics. Sampling with statistical analysis	$p < 1/100$	7
High	Likelihood that product would ship with defects. Sampling by sorting without statistical analysis	$p < 1/20$	8
High	Likelihood that product would ship with defects. Sampling by sorting without statistical analysis	$p < 1/10$	9
High	Likelihood that product would ship with defects. Item is not checked. Defect is latent.	$p > 1/10$	10

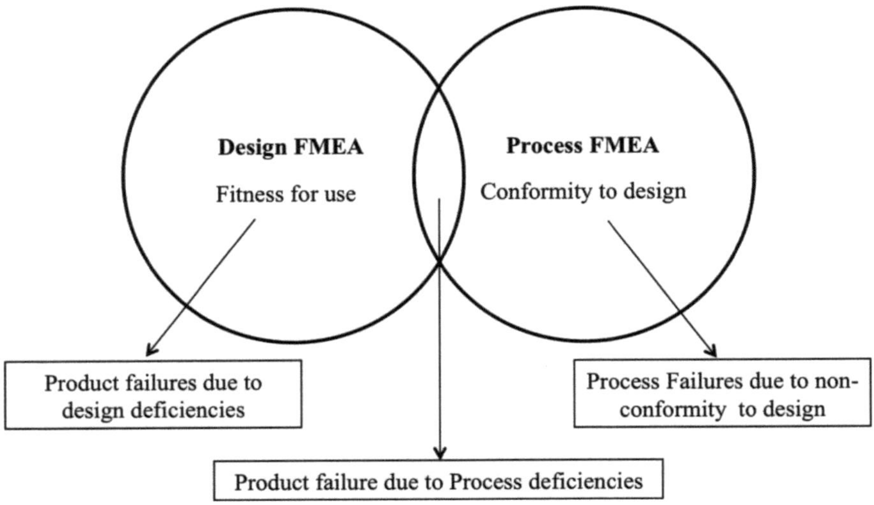

Fig. 5.9 Relationship between design FMEA and process FMEA. Modified from [10]

5.6.4 *Risk Priority Number*

After the previous analysis the team will be calculating the failure mode's risk priority number (*RPN*) as product of the severity (S), the occurrence (O), and the detection (D) ranking:

$$RPN = S \times O \times D \tag{5.17}$$

RPN is a measure of design risk. This value can assume values in the interval $1 \div 10 \times 10 \times 10 = 1{,}000$. This value (that could be represented as the volume of a parallelepiped) is used to rank the potential failures by their risks.

Remembering the Pareto Principle that says a high percentage of effects in any large system are caused by a low percentage of variables, and that *RPN* has no value in itself but must be consistent with other evaluation done by the same team. The team should establish which potential failure modes to address, conventionally establishing a guideline that could be the top one third or top one-half, Fig. 5.10. *RPN* derived by a severity or an occurrence rating of 9 or 10 are the first to be considered and actions should be directed to reduce them before any other analysis.

5.6.5 *Recommended Strategies*

Table 5.7 suggests the strategies that should be developed for reducing potential failures and driving down the risk. Table 5.8 offers a template for collecting the estimation of the three parameters, the calculation of the Risk Priority Numbers, the

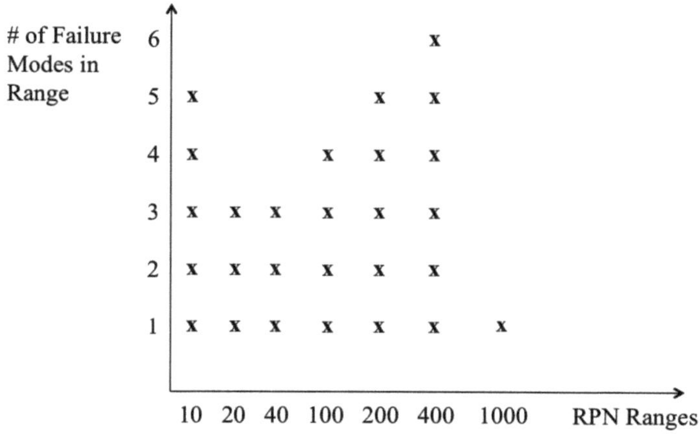

Fig. 5.10 Ranking the potential failures by *RPN*

Table 5.7 Strategies for a risk reduction

High occurrence rating	Attack the cause
High detection rating	Attack the controls
High severity rating	Plan a design review

Table 5.8 Chart for RPNs calculation and recommended actions

Product/ Process Description	Failure **Mode**	**Effects:** Severity	G	**Causes:** Occurrence	F	**Control:** Detectability	C	**RPN=** GxFxC	Depart./ Resp. Person	**Actions**	Resulting **RPN**
For every Function/ Component: 1.	List of Failure Modes: **1.1** **1.2** **1.3**	Description **2.1** **2.2** **2.3**									

actions to recommend to responsible persons and to address to the relative Department the suggestions of the team. Finally, after actions have been taken, the RPN must be recalculated and a new reduction of risk evaluated.

5.7 What We Have Learned from This Chapter

The Safety assessment was introduced as a necessary topic of the conceptual design of a product or a system.

The classic method of the safety factor, based on a deterministic evaluation, fixes safety margins: a system has a safety factor when the collapse is expected to occur with conditions markedly more rigorous than those provided as design requirements. Subsequently, various safety philosophies are discussed, suitable for different types of systems. The limits of the deterministic and the advantages of the probabilistic estimations are clarified. The last one best interprets the design variables that are random.

Finally, the FMEA method is illustrated which, with not only scientific but also logical and historical assumptions (based on the experience stored on previous designs), allows one to attain further levels of safety, comparing the risk of failure of different design solutions and suggesting actions for reducing it.

References

1. Allison (1990) FMEA: quality at source. In: Guidelines for using process FMEA, reliability engineering, Allison Transmission
2. Battazza M (2003) FMEA per la fabbricazione di un nuovo motore 125–180 cmc. Thesis a.a.2001/2002, University of Bologna, Engineering Faculty, IT, Relatore: A. Freddi
3. Birolini A (2017) Reliability engineering, 8th edn. Springer
4. Business Dictionary (2016) Concept. http://www.businessdictionary.com/definition/concept.html
5. Business Dictionary (2016) Team. http://www.businessdictionary.com/definition/team.html
6. Carlson CS (2012) Effective FMEAs. Wiley, New York
7. Connor A, Irizarry A (2015) Discussing design : improving communication and collaboration through critique. OReilly, Beijing
8. DOD (1980–1998) MIL-STD-1629A, military Standard: Procedures for performing a FMECA (24 NOV, 1980) Military Standard, Commanding Officer, Engineering Specifications and Standards Department (Code 93). Naval Air Engineering Center, Lakehurst, NJ, p 08733
9. Ford Motor Company (1995) Potential failure mode and effects analysis. Ford Environmental and Safety Engineering, Handbook
10. Freddi A (2004) Imparare a progettare, vol 1, 1st edn. Pitagora Editrice Bologna
11. Freddi A, Olmi G, Cristofolini L (2015) Experimental stress analysis for materials and structures. In: Stress analysis models for developing design methodologies. Series in solid and structural mechanics, vol 1, 1st edn. Springer
12. Larsson O (2012) Reliability analysis, Lund university. www.kstr.lth.se/fileadmin/kstr/pdf_files/vbkn10/.../Reliability_analysis_121101.pdf
13. Lidwell W, Holden K, Butler J (2003) Universal principles of design. Rockport Publication, 1
14. Mikulak RJ, McDermott R, Beauregard M (2008) The basics of FMEA, 2nd edn. CRC Press
15. O'Connor P (1991) Practical reliability engineering, vol 1, 3rd edn. Wiley
16. RAC-FMD-97CD (1997) Failure mode distribution data CD-ROM 1997. https://www.document-center.com/standards/show/RAC-FMD-97CD/history/1997%20EDITION
17. SAE J 1739 (1995) Potential failure mode and effects analysis. QS 9000 FMEA reference manual, SAE J 1739. https://www.lehigh.edu/~intribos/Resources/SAE_FMEA.pdf

18. Wikipedia (2015) FMECA. https://en.wikipedia.org/wiki/Failure_mode,_effects,_and_criticality_analysis
19. Wikipedia (2016) FMEA. https://en.wikipedia.org/wiki/Failure_mode_and_effects_analysis
20. Wikipedia (2017) de havilland comet. https://en.wikipedia.org/wiki/De_Havilland_Comet
21. Wikipedia (2017) Fail-safe. https://en.wikipedia.org/wiki/Fail-safe
22. Wirsching P, Ortiz K (1992) Reliability methods in mechanical and structural design. The University of Arizona, College of engineering, Aerospace & Mech. Dept., Tucson, Arizona 85721
23. WMG (2007) FMECA product excellence using Six Sigma. Warwick Manufacturing Group at School of Engineering University of Warwick, Coventry CV4 7AL, UK. http://www2.warwick.ac.uk/fac/sci/wmg/ftmsc/modules/modulelist/peuss/slides/section_12a_fmeca_notes.pdf

Chapter 6
Design of Experiment

Abstract In the case of lack of theoretical knowledge of the relationships between input and outputs variables of a system, we have a second possibility: the experimental reconstruction of these relationships. In the case of theoretical knowledge, the system is represented by linear or nonlinear equations between input and output variables, while in the absence of them an experiment must be planned to reach a certain level of knowledge of the system: the Design of experiment (DOE). This tool is general enough to be applied to different types of variables (e.g. categorical variables). Given the introductory character of this chapter, the presentation is limited to basic aspects.

6.1 The DOE Philosophy

Test and *experiment* have two different meanings. Test is an activity undertaken on a system in order to determine whether or not it meets the expectations in real operating conditions. If not, appropriate modifications must be introduced. Experiments, conversely, make it possible to check:

- Which variables (called factors) have a influence (effect) on the response of the system.
- For which variables values the system gives an optimal response.

Factors can be considered not constant, but varying at different levels, (as a minimum possibility at two levels, minimum and maximum level). Factors can be represented by continuous functions or discrete values.

We call *noise* the effect of all factors other than input. They are the most critical (but interesting) part of the experiment and the heart of the method. How can they be evaluated?

1. The first simplification is to argue that noises and their interactions with the main variables give no appreciable effects on response variables.
2. A more conscious attitude is to try to estimate their magnitude via earlier experiments.
3. Once known, they can be held constant (if it is possible) to avoid effects that manifest themselves in the form of instability of the output variables.

A. Freddi and M. Salmon, *Design Principles and Methodologies*, Springer Tracts in Mechanical Engineering, https://doi.org/10.1007/978-3-319-95342-7_6

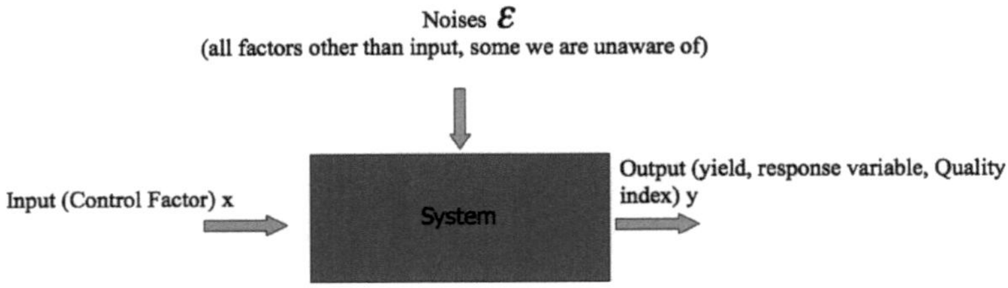

Fig. 6.1 Basic representation of a design of experiment. Modified from [10]

4. Prudently it's as well to randomize the experiments, i.e. randomly allocate them, in order to reduce bias, by equalizing the factors that have not been explicitly accounted for in the experimental design.
5. The best practice is to estimate the magnitude of noise effects by replicating the experiments.

6.1.1 Analysis of Variance of One Factor Design

A general scheme of the simplest Design of Experiment is shown in Fig. 6.1.

In the case of single input and output variables x and y, the objective is to determine the influence of independent variable x on dependent variable y. The mathematical (unknown) model is:

$$y = f(x, \epsilon)$$

where ϵ is the impact of all factors other than x (some unknown), i.e. of the noise.

Example 6.1 (*Reduction of weight due to a drug* [3])
The Table 6.1 shows the following data:

- In the first row the percentage of drug x (i.e. the factor levels (8).
- In the first column the repetitions number (3).
- In columns the test replications.
- Inside the boxes the output y values, given in pounds (*lb*).
- In the last row the means of the three repetitions for each column, with the symbol $\overline{y}_{.j}$.

The question is:

How does % drug impacts on weight?

Figure 6.2 illustrates the distribution of the data of Table 6.1 with indication of the replications numbers $i = 1, \ldots, 3$ and the mean values of each column (called with-in-mean of each column).

The statistical model is:

$$y_{ij} = \mu + \tau_j + \epsilon_{ij} \tag{6.1}$$

Table 6.1 Weight reduction through a drug, at different % of the active chemical ingredient, taken from [3]

No.	1%	2	3	4	5	6	7	8
1	10	8	27	23	16	16	25	28
2	18	19	17	18	25	8	23	24
3	8	16	16	28	22	19	30	20
$\overline{y}_{.j}$	12	17	20	23	21	17	26	24

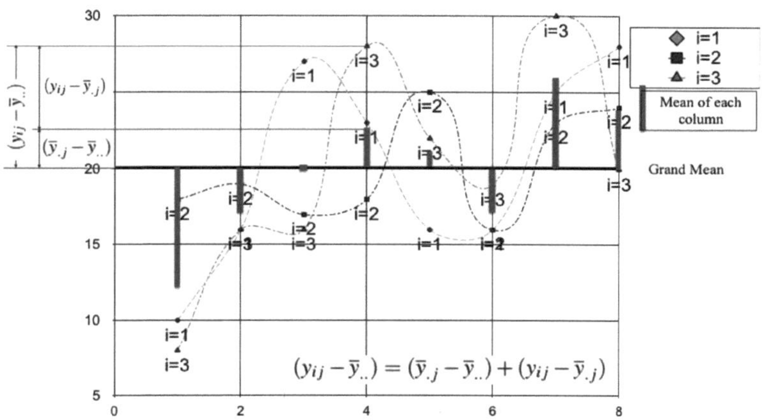

Fig. 6.2 Results of the experiment. (*Note* The values are joined by a continuous line only to show their fluctuation, but without physical meaning)

where:

- $i = 1, \ldots, R$ are the replications numbers.
- $j = 1, \ldots, C$ factors levels.
- μ = overall average.
- τ_j differential effect associated with j-th level of x. Note that $\left(\sum_{j=1}^{C} \tau_{ij} = 0\right)$.
- ϵ_{ij} noise or error associated with any particular data value (i, j).

The experiment gives a total of $R \cdot C$ values of y_{ij}

From a data sample it is possible to have estimates of μ, τ_1, τ_2, τ_C then, as difference, also the estimate of the error ϵ_{ij}

$\overline{y}_{..}$ is an estimate of μ the true mean (that is unknown).

$\overline{y}_{.j} - \overline{y}_{..}$ is an estimate of τ_j

$\hat{\epsilon}_{ij} = y_{ij} - \overline{y}_{.j}$ as estimation of the error

We insert these estimates in the statistical model (in virtue of Gauss principle of the least squares):

$$(y_{ij} - \overline{y}_{..}) = (\overline{y}_{.j} - \overline{y}_{..}) + (y_{ij} - \overline{y}_{.j}) \quad (6.2)$$

This equation can be read as:

Total variability in y is equal to variability in y due to x plus variability in y due to all other factors.

If we square both sides and double sum both sides, after some simplifications, we get the following expression:

$$\sum_{j=1}^{C}\sum_{i=1}^{R}(y_{ij}-\bar{y}_{..})^2 = R\sum_{j=1}^{C}(\bar{y}_{.j}-\bar{y}_{..})^2 + \sum_{j=1}^{C}\sum_{i=1}^{R}(y_{ij}-\bar{y}_{.j})^2 \quad (6.3)$$

This way of dealing with the problem with quadratic sums reveals an interesting property:

The total variation with respect to the general mean (in this case the grand mean) is equal to the sum of squares between columns (between tests) and the sum of squares within columns (within tests).

$$TSS_{(total)} = SSB_{c\,(between\ columns)} + SSW_{(within\ columns)}$$

where:

TSS Total Sum of Squares
SSB_{Col} Sum of Squares between columns
SSW_{error} Sum of squares within columns.

6.1.1.1 Analysis of Variance (ANOVA)

The following Table 6.2 formalizes the analysis of variance, called ANOVA, i.e. shows the significance of all contributions, compared to the total variance of the results. The second column of Table 6.2 contains the sums of squares previously

Table 6.2 Analysis of variance (ANOVA)

Source of variation	Sum of squares	df	Mean square	F_{calc}
Between columns SSB_c	$R\sum_{j=1}^{C}(\bar{y}_{.j}-\bar{y}_{..})^2$	$C-1$	$MSB_c = \frac{SSB_C}{C-1} = V_{col}+\sigma^2$	$\frac{MSB_c}{MSW}$
Within columns SSW	$\sum_{j=1}^{C}\sum_{i=1}^{R}(y_{ij}-\bar{y}_{.j})^2$	$C(R-1)$	$MSW = \frac{SSW}{(R-1)C} = \sigma^2$	–
Total TSS	$\sum_{j}^{C}\sum_{i}^{R}(y_{ij}-\bar{y}_{..})^2$	$RC-1$	–	–

Table 6.3 Numerical data of the first ANOVA

Cause of variation	Sum of squares	df	Mean square	F_{calc}
SSB_C	432	7	61.7	$61.7/18.30 = 3.37$
SSW error	292	16	18.3	
TSS total	724	23		

defined. The third column shows the degrees of freedom of every sum. The ratios between the sums and the degrees of freedom allow comparisons otherwise impossible between the various contributions [16]. Regarding the degrees of freedom:

- Number of all possible comparisons between treatments combinations (Columns) is $C - 1$
- Number of all possible comparisons between replications (Rows) i.e. within each treatment combination is $R - 1$. Then for all treatment combinations: $C(R - 1)$.
- Number of all possible comparisons between every RC values and the *Grand Mean* is $RC - 1$.

The numerical values of all these quantities for the example of Fig. 6.2 are shown in the Table 6.3. where:

- The sum of squares is $SSB_c = 432$. If it is divided by $\nu_1 = 7 = (8 - 1)$ (the degrees of freedom between the 8 mean values of the 8 columns), the mean Square value between the columns is $MSB_c = 432/7 = 61.71$.
- With the same argument we can calculate the Mean Square for the data within columns, observing that the comparisons between the data in each row within each column are 2, (3 being the number of repetitions) but the number of columns is 8. So the total degrees of freedom are 2×8 i.e. $\nu_2 = 16$. Then $MSW = 292/16 = 18.25$.[1]
- If $MSB_c/MSW \leq 1$ there is no evidence that $V_{col} > 0$, i.e. that the level of x affects y.
- If $MSB_c/MSW > 1$ there is some evidence of no-zero V_{col} or level of x affects y.

6.1.2 Note on the Hypothesis Testing

We can give more scientific basis to this evidence (governed by common sense), observing that this ratio $F_{calc} = MSB_c/MSW$ follows a statistical law.

If the column means are equal, and the distribution of the population is gaussian, the ratio has a *Fisher distribution* with $\nu_1 = C - 1$ and $\nu_2 = (R - 1)C$ degrees of freedom, Fig. 6.3.

[1] $E(MSW)$ is the expected value MSW for an infinite number of repetitions. Assuming the variance σ^2 for each value y_{ij} as constant, we can assume that: $E(MSB_c) = \sigma^2 + V_{col}$. ($V_{col}$ being the measure of differences among column means.)

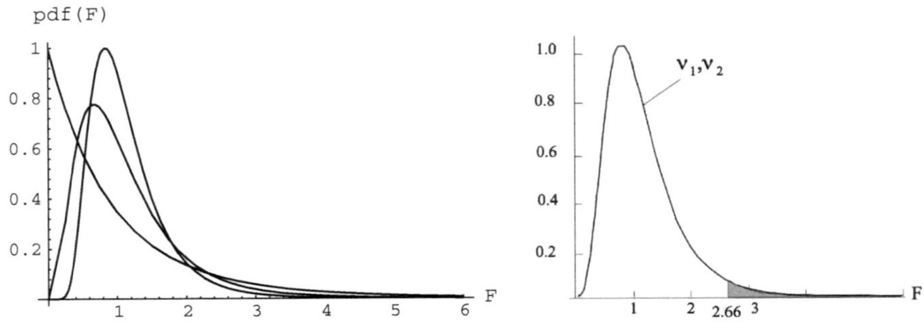

Fig. 6.3 Fisher probability density functions for different degrees of freedom. On the right the distribution for ν_1 and ν_2 degrees of freedom

We compare two hypotheses called H_0 and H_1.

- Hypothesis H_0 is verified when the column means do not have significant differences (i.e. they are statistically equal: $(\mu_1 = \mu_j = \mu_c)$; then $V_{col} = 0$. This means that levels of x have not impact on y.
- Hypothesis H_1 is verified when levels of x do have an impact on y. This happens when
$$\frac{MSB_c}{MSW} >> 1$$
In our Example 6.1 and in Table 6.3, $MSB/MSW = 3.37 >> 1$. Then we can reject hypothesis H_0.

Practically, we operate in this way:

1. We fix a small critical threshold of probability value (e.g. $\alpha = 0.05$). This threshold is a free choice of the experimenter, i.e. *it must be pre-set*. It is represented by the gray area in Fig. 6.3 and it establishes the level of confidence the experimenter wants, for acceptance of the conclusions.
2. We calculate the critical value of F (C) corresponding to this threshold (in the Example 6.1, for $\nu_1 = 7$ and $\nu_2 = 16$, the critical value is $C = 2.66$). If $F_{calc} > C$, the probability that the hypothesis H_0 is true is less than 5%. *Consequently, we assume the hypothesis H_1 as true.*
3. Then α is the probability of rejecting the hypothesis H_0 when H_0 is true, i.e. it is the Type *I* error:
$$\alpha = P(type\ I\ error) = P(reject\ H_0,\ \ when\ H_0\ is\ true)$$
4. If vice-versa, $F_{calc} \leq C$ we accept H_0 (null hypothesis) and reject the hypothesis H_1 (alternate hypothesis). (To accept H_0 means rejecting H_1, while rejecting H_0 means accepting H_1).
5. We define β the probability of type II error, that we must consider when $F_{calc} < C$. It is:

Table 6.4 Remarks on errors

	H_0 True	H_0 False
Accept H_0	No error	Type II error
	$1-\alpha$	β
Reject H_0	Type I error	No error
	α	$1-\beta$

$$\beta = P(Type\ II\ error) = P(accept\ H_0,\ \ when\ H_0\ is\ false).$$

6. $1-\beta$ (called Power of the Test) is then the probability to reject H_0, when H_0 is false, i.e. a correct conclusion Table 6.4.

As concluding remarks, the hypothesis H_0 must be accepted only if the results are *extraordinarily evident*, (that is not the case of the Example 6.1).

6.1.2.1 Graphical Representation of ANOVA

A graphical representation of data of Example 6.1 illustrates the different level of "clustering" of (Fig. 6.4):

- Total data around the grand mean.
- Columns means around the grand mean.
- Data within each column around their column mean.

The *first image* shows a dispersion that does not encourage any conclusion, the *second image* shows a certain efficacy of diet over the general average, and the *third* shows how the experimental error is sufficiently contained to allow rejection of the hypothesis H_0 even if the evidence is not overwhelming.

It is worth mentioning at this point that all input factors and all responses must be measured to an appropriate degree of *accuracy and precision.* If they are not, the measurement error will mask the actual variation due to different settings of the factors themselves [14].

According to ISO 5725-1, Accuracy consists of Trueness (proximity of measurement results to the true value) and Precision (repeatability or reproducibility of the measurement.)

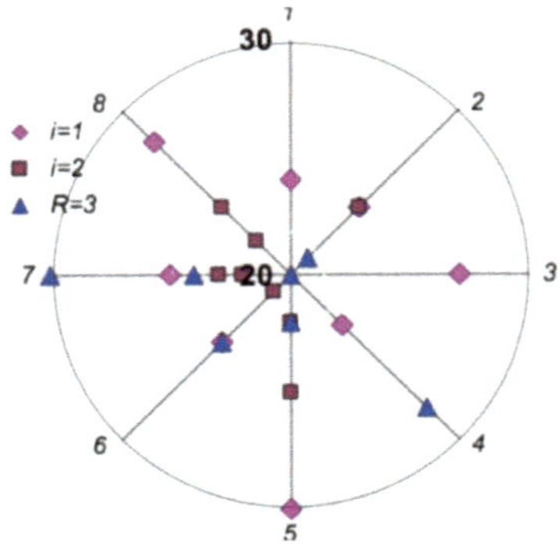

(a) *Distribution of data i, j with respect to the grand mean,*

$$TSS = \sum_{j}^{C}\sum_{i}^{R}(y_{ij} - \overline{y}_{..})^2 = 724$$

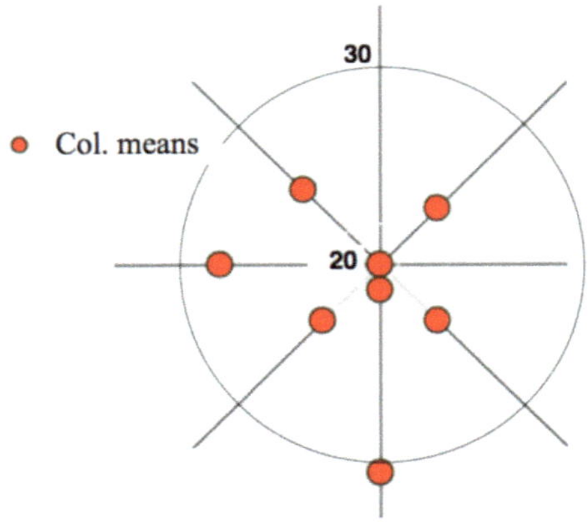

(b) *Distribution of the columns means with respect to the grand mean,*

$$SSB = \sum_{j=1}^{C}\sum_{i=1}^{R}(y_{ij} - \overline{y}_{.j})^2 = 432$$

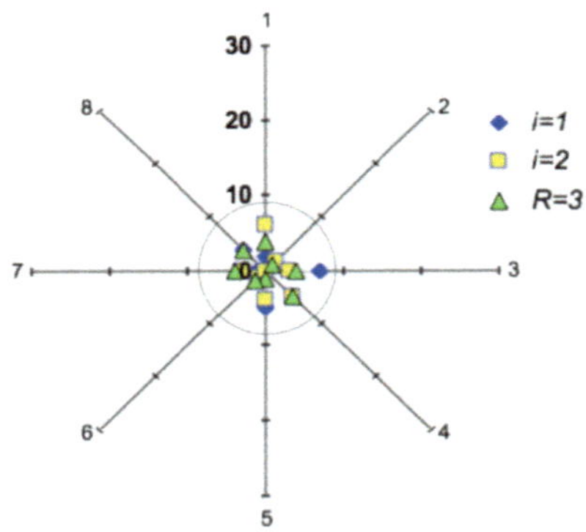

(c) *Distribution of data i, j in each column with respect to the corresponding column mean,*

$$SSW = \sum_{j=1}^{C}\sum_{i=1}^{R}(y_{ij} - \overline{y}_{.j})^2 = 292$$

Fig. 6.4 Graphical representation of the distributions of the data with respect to their means values. Modified from [10]

Continuous variables are present when any dependent variable can be measured on a continuous interval or ratio scale through a *measuring device*. Data can be interval and ratio data, i.e. data as time, speed, distance and temperature that can be measured on interval scales and we have proper tools to do it. They are called interval scales because the differences between the consecutive numbers are of equal intervals. We can calculate mean, variance and other statistics and we can perform *a parametric test*, with these assumptions:

- ϵ_{ij} are independent random variables. $E(\epsilon_{ij}) = 0$ is the expected value for a infinite number of data.
- Variance $\sigma^2(\epsilon_i j)$= constant for all values i, j.
- Each ϵ_{ij} is normally distributed.

A different logic has to be applied to the data to make decisions about the samples and the populations that do not require the assumptions of a parametric test. We speak of *non-parametric tests*.

- In non-parametric tests the dependent variable is *ordinal* and a non-parametric analysis can be developed based only on the rank ordering of data.
- *Ordinal data* define only an order of performance. So we can have a rank order only and we cannot calculate means and variance on ordinal data.
- *Nominal data* are another type for defining different categories. It is meaningless to calculate statistics on these. But we can have numerical data as data frequencies in each category. In this case the numbers indicate the category the subjects belong to instead of a position in an ordinal scale or a value in an interval scale. So we can collect and treat frequency data, adding up the number of subjects in each category.
  The typical statistic of this case is the chi-square statistic that allows us to analyze frequency of data and to answer this question: the frequency distribution measured in a sample is really significant of a difference between the categories or is it merely casual?

6.1.3 General Expression of Designs with Two Factors at Multiple Levels and Replications

We start with two factors design in the case of R levels for the first variable, C levels for the second variable and k replications.

The mathematical model of a design of this type is:

$$y_{ijk} = \mu + \rho_i + \tau_j + I_{ij} + \epsilon_{ijk} \tag{6.4}$$

with:
$i = 1, \ldots, R$ levels of rows
$j = 1, \ldots, C$ levels of columns
$k = 1, \ldots, n$ number of replications
I_{ij} = Interaction effect (it is a degree of difference from the sum of separate effects).

The approximate model is:

$$y_{ijk} = \overline{y}_{...} + (\overline{y}_{i..} - \overline{y}_{...}) + (\overline{y}_{.j.} - \overline{y}_{...}) + (\overline{y}_{ij.} - \overline{y}_{i..} - \overline{y}_{.j.} - \overline{y}_{...}) + (y_{ijk} - \overline{y}_{ij.}) \quad (6.5)$$

$\overline{y}_{...}$ = grand mean
$\overline{y}_{i..}$ = mean of row i
$\overline{y}_{.j.}$ = mean of column j
$\overline{y}_{ij.}$ = mean of box ij

The first term of Eq. 6.5 is the mean, the second is row effect, the third is column effect, the fourth is interaction effect and the last one is error.

Squaring and summing three times over i, j and k we get, after all cross-product terms are cancelled:

$$\sum_{i}^{R}\sum_{j}^{C}\sum_{k}^{n}(y_{ijk} - \overline{y}_{...})^2 = n \cdot C \cdot \sum_{i}^{R}(\overline{y}_{i..} - \overline{y}_{...})^2 + n \cdot R \cdot \sum_{j}^{C}(\overline{y}_{.j.} - \overline{y}_{...})^2 +$$

$$+ n \cdot \sum_{i}^{R}\sum_{j}^{C}(\overline{y}_{ij.} - \overline{y}_{i..} - \overline{y}_{.j.} + \overline{y}_{...})^2 + \sum_{i}^{R}\sum_{j}^{C}\sum_{k}^{n}(y_{ijk} - \overline{y}_{ij.})^2$$

or:

$$TSS = SSB_{Rows} + SSB_{Col} + SSI_{R,C} + SSW_{error}$$

where:

TSS Total Sum of Squares
SSB_{Rows} Sum of Squares between rows
SSB_{Col} Sum of Squares between columns
$SSI_{R,C}$ Sum of Squares of interactions between factors
SSW_{error} Sum of squares within columns.

R is the number of rows, C is the number of columns and n is the number of replications. The following Table gives the general scheme for ANOVA, Table 6.5. For the degrees of freedom the following relationship is valid:

$$R \cdot C \cdot n - 1 = (C - 1) + (R - 1) + (R - 1)(C - 1) + RC(n - 1)$$

Table 6.5 ANOVA for a designs with two variables at multiple levels and one replication

Source of variation	Sum of squares	*df*	Mean square (MS)
Rows	$SSB_{row} = n \cdot C \cdot \sum_{i}^{R} (\bar{y}_{i..} - \bar{y}_{...})^2$	$C - 1$	$\frac{SSB_R}{C-1}$
Columns	$SSB_{col.} = n \cdot R \cdot \sum_{j}^{C} (\bar{y}_{.j.} - \bar{y}_{...})^2$	$R - 1$	$\frac{SSB_C}{R-1}$
Interact.	$SSI_{R,C} = n \cdot \sum_{i}^{R} \sum_{j}^{C} (\bar{y}_{ij.} - \bar{y}_{i..} - \bar{y}_{.j.} + \bar{y}_{...})^2$	$(R-1)(C-1)$	$\frac{SSI_{R,C}}{(R-1)(C-1)}$
Error	$SSW_{error} = \sum_{i}^{R} \sum_{j}^{C} \sum_{k}^{n} (y_{ijk} - \bar{y}_{ij.})^2$	$RC(n-1)$	–
Total	$\sum_{i}^{R} \sum_{j}^{C} \sum_{k}^{n} (y_{ijk} - \bar{y}_{...})^2$	$RC \cdot n - 1$	

where:
$R \cdot C \cdot n$ are the total number of points
$(R-1)(C-1)$ are the *df* of interactions, always equal to the product of the df of rows and columns.

6.2 Design with Two Variables at Two Levels and One Replication

The main advantage of factorial designs is the possibility to examine the effects of several factors simultaneously, called interaction. It is the phenomenon, upon which, the effect of one factor depends on the level of another factor.

Two factors at two levels $2 \times 2 = 4$ give the minimum chance to perform this analysis.

With this design, each factor has two fixed levels. One level is labeled low and the other high in an arbitrary way. Levels may be quantitative or qualitative, with four possible experimental combinations, called treatments.

We assume that an experiment (with replications) is performed to evaluate the effects estimates of two factors x_1 and x_2 at two levels (low and high), called $-1, +1$ for the first and $-1, +1$ for the second, on a single output variable y.

Figure 6.5a, b shows two examples of absence and presence of interaction.

The values of normalized input variables are shown in Fig. 6.6 together with the output variables with the so-called *Yates notation*.

N	x_1	x_2	y_1	y_2	y_m
1	-1	-1	5	6	5.5
2	+1	-1	10	9	9.5
3	-1	+1	8	9	8.5
4	+1	+1	12	13	12.5

(a) *Design matrix without interaction*

N	x_1	x_2	y_1	y_2	y_m
1	-1	-1	5	6	5.5
2	+1	-1	10	9	9.5
3	-1	+1	8	9	8.5
4	+1	+1	17	18	17.5

(b) *design matrix with interaction*

Fig. 6.5 Example of 2×2 design with a replication without and with interaction ($C = R = n = 2$). The output values are due to two treatments: y_1 first treatment, y_2 a replication, $\overline{y}_i = y_m$ the average value

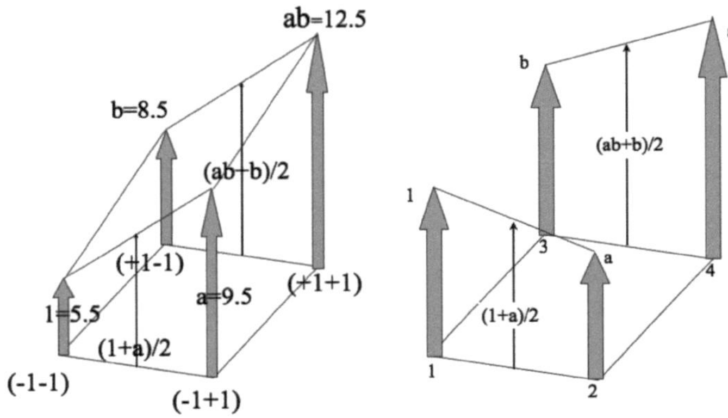

Fig. 6.6 Design with two factors and two levels with Yates notation for response variable. On the left without interaction and on the right with positive interaction

- $-1\,-1$ = minimum level of both variables. Response variable assumes the notation 1
- $+1\ -1$ = max of the first, min of the second. Response variable assumes the notation a
- $-1\ +1$ = min of the first, max of the second. Response variable assumes the notation b
- $+1\ +1$ = max of both. Response variable assume the notation ab.

If $ab = 12.5$ no interaction is present, while if $ab > 12.5$ there is a positive interaction and on the contrary, if $ab < 12.5$ there is negative interaction.

A graphic representation of Fig. 6.6 clarifies the meaning of interaction: if the four points of ordinates $(1, a, b, ab)$ belong to a plane, there is no interaction (see Fig. 6.6 on the left side). If, conversely, the segments that ideally join the points are not parallel two to two (see Fig. 6.6 on the right side), they cannot locate a plane but only a warped surface, i.e. a doubly ruled surface, a symptom of interaction. We can see the same effects in Fig. 6.7, where that shows the same conclusion in another way.

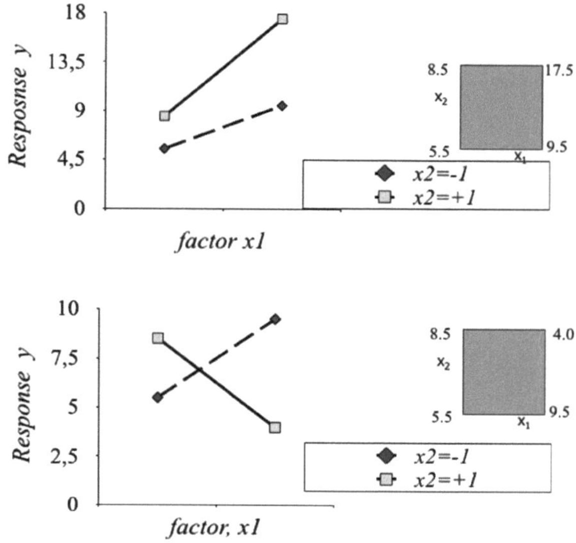

Fig. 6.7 A graphical representation of a positive and of a negative interaction

The 2^n refers to designs with n factors where each factor has just two levels. They are a major set of building blocks for many experimental designs. These designs are usually referred to as screening designs, to explore a large number of factors, each having just two levels [1].

We point out that the design of two variables at two levels and only one dependant variable (yield) can cover a variety of cases of real interest to scientists and engineers. It can be useful in the optimization of a technological process as well in the design of a system.

One of the advantages is that, while both factors vary simultaneously, the three effects estimates (A, B, AB) are determined separately.

On the other hands, a preliminary identifying of presumed control factors and fixing of their levels is a delicate pre-work that can benefit from tools like the Ishikawa fish-bone diagram, the House of Quality and other means that help to hypothesize the sources of normal or abnormal effects.

By means of the expressions of Table 6.5 the data of Fig. 6.5b, d, (for the mean values of y_i and two replications), produce the ANOVA table in Table 6.6.

The effects calculation is done for the means values, so that it can be visualized on the mean planes [6]. E.g. the Mean Effect of the first variable x_1 is $A = (a + ab)/2 - (1 + b)/2 = 13.57 = 6.5$, shown in Fig. 6.8. The effects estimates are identified according to the following notation:

- $i = 1, \ldots, m$ Number of T.C.'s
- $j = 1, \ldots, n$ Number of replications
- $1, a, b, ab$ Yates notation

Table 6.6 ANOVA for a design with two factors at two levels

Source of variation	Sum of squares	*df*	Mean square	F_{calc}
Rows	84.5	1	84.5	169
Columns	60.5	1	60.5	121
Interactions	12.5	1	12.5	25
Error	2.00	4	$\sigma^2 = 0.5$	
Total	159.5	7		

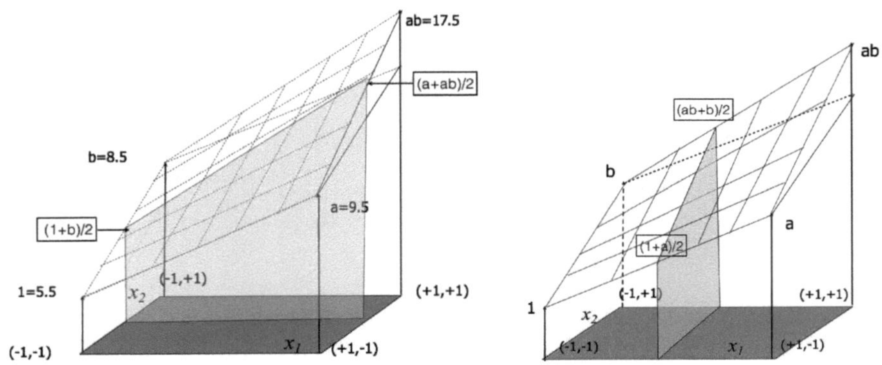

Fig. 6.8 First and second effect calculation: a graphical representation, according to the graphical suggestion in [6]

- A = effect on y of the first variable x_1
- B = effect on y of the first variable x_2
- AB = combined effect on y of the two variables $x_1 x_2$.

For $m = 4$ T.C.'s, and two replications ($n = 2$) we get the following expressions, Fig. 6.8.

$$A = \frac{a + ab}{2} - \frac{1 + b}{2} \quad B = \frac{b + ab}{2} - \frac{1 + a}{2} \quad AB = \frac{ab - b}{2} - \frac{a - 1}{2} \tag{6.6}$$

All the effects estimates of variables x_1 and x_2 are computed faster (for the **mean** $\overline{y}_i$ value of the n replications of the answer), with the help of a matrix, called *calculation matrix*, Table 6.7.

The calculation matrix has the following properties:

- The normalized values of the independent variables (x_1, x_2) are shown only through their signs ($x_1 = +$ means $x_1 = +1$ etc.).
- The calculation matrix is an *orthogonal* array. In an orthogonal array the columns are such that each level of a factor crosses all levels of the other factors and the sum of the products of the corresponding variables of two columns is always zero.

Table 6.7 Calculation matrix– $\overline{y}_{i.} = (\sum_{j=1}^{n} y_j)/2$ = mean between replications

T.C.'s	I	x_1	x_2	x_1x_2	y	y_1	y_2	$\overline{y}_i$
$i = 1$	+	−	−	+	1	5	6	5.5
2	+	+	−	−	a	10	9	9.5
3	+	−	+	−	b	8	9	8.5
m = 4	+	+	+	+	ab	17	18	17.5
	$\sum/4$	A = $\sum/2$	B=$\sum/2$	AB=$\sum/2$				
	10.25	6.5	5.5	2.5				

- The signs of term x_1x_2 (obtained multiplying the corresponding values of x_1 and x_2), is used to calculate the interaction.
- The effects estimate calculation is performed in the bottom row of the matrix for $j = 4$ replications with respect to the mean value $\overline{y}_{i.}$, with $i = 2$, according to the following algorithm: A, B, AB are obtained multiplying the output values times the columns variables (x_1 etc.) and dividing the sum of this multiplications by half of the number of treatment combinations (4/2).

$$A = \frac{-1 + a - b + ab}{2} = 6.5 \quad B = \frac{-1 - a + b + ab}{2} = 5.5$$

$$AB = \frac{1 - a - b + ab}{2} = 2.5$$

- The effects number (A, B, AB) is always one less than the treatments number (four in this case).
- The first column gives the grand mean value, obtained dividing the sum of the output values by the number of replications.

6.2.1 *Response Surface for the Design with Two Variables at Two Levels*

When the one or both factors are *qualitative* (such as, e.g. the presence or not of a chemical additive with low level one type and the high level another type) and the effects estimates are determined, the combination that gives the best results does not need further development.

If the variables are continuous, the yield can be interpolated by a continuous response surface. This offers the advantage of a complete view of the system response and allows estimation even at points other than those directly considered in the experiment.

The simplest interpolating surface for four points not positioned on one plane (that happens in the case of interaction), is a *ruled surface* that, for two factors, is described by an equation like this:

$$y = b_0 + b_1 x_1 + b_2 x_2 + b_{12} x_1 x_2 + \epsilon \tag{6.7}$$

In the case of replications, the surface can be calculated for the mean values of several outputs. Referring to the example of data in Table 6.7, it is easy to recognize that the coefficients of the previous equation are the effects estimates divided by 2, the effects estimates being the increment of response for two units of variations of the input variables from -1 to 0 and from 0 to $+1$. With this observation, the response surface equation as a function of the effects estimates, becomes:

$$y = b_0 + \frac{A}{2} x_1 + \frac{B}{2} x_2 + \frac{AB}{2} x_1 x_2 + \epsilon \tag{6.8}$$

The additional term ϵ remembers that the output is also an estimation. For the previous example:

$$\hat{y}_i = 10.25 + \frac{6.50}{2} x_{i1} + \frac{5.50}{2} x_{i2} + \frac{2.50}{2} x_{i1} x_{i2} \tag{6.9}$$

where $\hat{y}_i$ is an estimation of y_i, based on effects obtained by the mean value of two tests.

6.3 Design with Three Factors at Two Levels

For three dimensions the design domain is a cube and with four or more factors a hypercube.

Also in this case the design aim is to find the tern of input variables that gives the optimal output. A standard method uses four tests for 4 domain points.

- $-1, -1, -1$ (point 1)
- $+1, -1, -1$ (point 2)
- $-1, +1, -1$ (point 3)
- $-1, -1, +1$ (point 5)

With four tests only, each variable would be utilized once at its maximum level and three times at its minimum level, Fig. 6.9, (on the left side). On the contrary, information is complete only with eight tests, Fig. 6.9 (on the right side), where all the variables and all the interactions are considered.

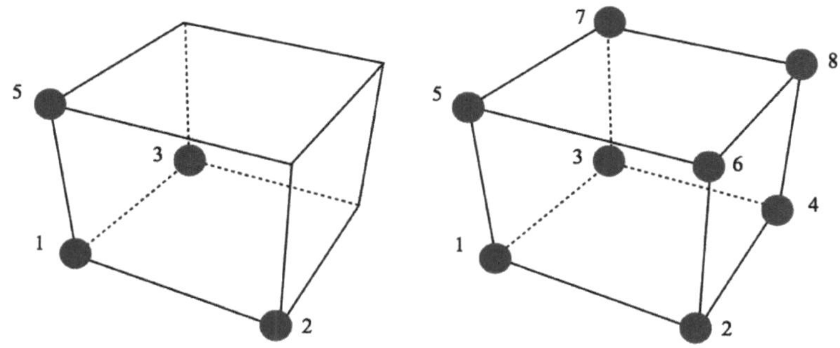

Fig. 6.9 Definition domain of a three factors design

Table 6.8 Calculation matrix for a 2^3 design. Modified from [10]

N	I	x_1	x_2	x_1x_2	x_3	x_1x_3	x_2x_3	$x_1x_2x_3$	y
1	+	-	-	+	-	+	+	-	1
2	+	+	-	-	-	-	+	+	a
3	+	-	+	-	-	+	-	+	b
4	+	+	+	+	-	-	-	-	ab
5	+	-	-	+	+	-	-	+	c
6	+	+	-	-	+	+	-	-	ac
7	+	-	+	-	+	-	+	-	bc
8	+	+	+	+	+	+	+	+	abc
Effects	Σ/8	A= Σ/4	B= Σ/4	AB= Σ/4	C= Σ/4	AC= Σ/4	BC= Σ/4	ABC= Σ/4	divided by N/2
Coeffic. b	b_0=Σ/8	b_1= A/2	b_2= B/2	b_{12}= AB/2	b_3= C/2	b_{13}= AC/2	b_{23}= BC/2	b_{123}= ABC/2	

6.3.1 *Calculation Matrix (Three Factors at Two Levels)*

The matrix for calculation of all effects estimates of a 2^3 design is shown in Table 6.8 with the same meaning as the previous Table 6.7 ($-$ and $+$ stay for -1 and $+1$ respectively), and with the same algorithm to calculate all the coefficients. In this table:

- The first column is the test number.
- The second is the I column and allows computation of the mean b_0 as the sum of the values y of the last column (all with sign $+$) divided by 8.

- The normalized values of independent variables (3 factors x_1, x_2, x_3) are in the next three columns, where:
  -1 is the low level
  $+1$ is the high level
  of the three factors.
- The values in the remaining columns are obtained multiplying row by row the previous ± 1 values.

According to usual notation, 1 means the y value for the three factors at their minimum level, $-1, -1, -1$, a means the y value for the first factor at the maximum level $+1$, and the second and third factors at their minimum levels $-1, -1$ and similarly for the others.

A comparison with the intuitive but limited evaluation of four points only shows that we get the main and interaction effects estimate, multiplying the yield values in the $\overline{y}$ column by the corresponding values $-1, +1$ in the columns. Each product must be algebraically summed and finally divided by half of the number of treatments (in this case $2^k/2 = 2^3/2 = 4$):

$$A = \frac{-1 + a - b + ab - c + ac - bc + abc}{4} \tag{6.10}$$

$$B = \frac{-1 - a + b + ab - c - ac + bc + abc}{4} \tag{6.11}$$

$$C = \frac{-1 - a - b - ab + c + ac + bc + abc}{4} \tag{6.12}$$

$$AB = \frac{+1 - a - b + ab + c - ac - bc + abc}{4} \tag{6.13}$$

$$AC = \frac{+1 - a + b - ab - c + ac - bc + abc}{4} \tag{6.14}$$

$$BC = \frac{+1 + a - b - ab - c - ac + bc + abc}{4} \tag{6.15}$$

$$ABC = \frac{-1 + a + b - ab + c - ac - bc + abc}{4} \tag{6.16}$$

6.3.2 Response Surface (Three Factors at Two Levels)

The effects estimates, that are 7 (one less than the treatments number A, B, AB, C, AC, BC, ABC), are obtained by the calculation matrix of Table 6.8. Also in this case the simplest general mathematical model is given by a multi-linear function in three variables:

$$y = b_0 + b_1x_1 + b_2x_2 + b_3x_3 + b_{12}x_1x_2 + b_{13}x_1x_3 + b_{23}x_2x_3 + b_{123}x_1x_2x_3 + \epsilon \tag{6.17}$$

For a 2^k factorial design there is a model coefficient corresponding to each of the main and interaction effects, plus the mean response b_0. Estimates of the model coefficients are reported in the last row and may be obtained directly from the effects estimates. If the coefficients estimates are inserted into the previous equation, we get:

$$\hat{y} = \hat{b}_0 + \frac{A}{2}x_1 + \frac{B}{2}x_2 + \frac{C}{2}x_3 + \frac{AB}{2}x_1x_2 + \frac{AC}{2}x_1x_3 + \frac{BC}{2}x_2x_3 + \frac{ABC}{2}x_1x_2x_3 \tag{6.18}$$

6.4 Three or More Levels

As the factors number increases, there is a rapid increase in the measures that need to be performed: an experiment with 2 factors at k levels requires k^2 observations, while a design with 3 factors at k levels requires k^3 observations and so on (e. g. a 3-factor experiment, each with 4 levels, requires $4^3 = 64$ tests): time commitment and resources tend to unacceptable values.

6.4.1 Response Surface (Two Factors at Three Levels)

For the three level design, with eight tests on the boundary, it is possible to utilize a mathematical expression of response surface with a second-order function (so as to be able to follow the curvature of the response). The mathematical general expression is:

$$\hat{y} = b_0 + b_1x_1 + b_2x_2 + b_3x_1^2 + b_{12}x_1x_2 + b_4x_2^2 \tag{6.19}$$

Two level designs means two levels of each factor: using a 2^k design, even with a simple response surface we get limited information.

The reason for three-level designs is to have more information on intermediate points on the domain borders and to model possible curvature in the response function, in spite of higher experimentation cost. Three levels conventionally refer to low, intermediate and high levels. We can introduce the following notation for x_i values $-1, 0, +1$.[2]

The question that we ask is the following:

if the number of experiments is reduced with respect to their theoretical number, can we still get an adequate representation of the relationship between output and input? [8, 18].

---

[2]This may be confusing with respect to the 2-level designs since 0 seems reserved for central points between -1 and $+1$. Therefore, in three levels matrices we will use 1, 2, 3 notation that does not necessarily mean that 2 is the midpoint between 1 and 3.

6.5 Hints of Fractional Factorial Design

We explore the possibility to develop a method that makes it possible to simultaneously analyze multiple factors with a reduced number of data [8, 20]. We make a very small mention below of the much more complex problem of *confounding* and of fractional factorial design. This introduction with an example is, however, sufficient for our purpose. The definition of The ASQC (1983) Glossary and Tables for Statistical Quality Control is the following:

> Fractional design is a factorial experiment in which only an adequately chosen fraction of the treatment combinations, required for the complete factorial experiment, is selected to be run.

Consider a 2^k experiment with the idea of running fewer than 2^k t.c.'s.

Example 6.2 (*Fractional factorial design with three variables at two levels but with only four runs*)
Let's show the example of a full factorial design $2^3 = 8$ but with this constraint: we do not have enough resources to run more than $2^{3-1} = 4$ t.c.'s. The reduced number of treatment combinations (e.g. 1, a, b, c, abc) does not make it possible to determine all 7 effects (A, B, C, AB, AC, BC, ABC).

In fact, to compute A we need *all* the treatments combinations:

$$4A = -1 + a - b + ab - c + ac - bc + abc$$

and the same for BC:

$$4BC = 1 + a - b - ab - c - ac + bc + abc$$

But if we sum the two expressions we get:

$$4A + 4BC = 2(a - b - c + abc)$$

and in the same way:

$$2(A + BC) = a - b - c + abc$$

$$2(B + AC) = -a + b - c + abc$$

$$2(C + AB) = -a - b + c + abc$$

Then these *linear combinations of effects* were obtained by only 4 t.c.'s instead of 8.

Resuming, with 4 tests instead of 8, we measure only,
a, b, c, abc
and not:
$1, ab, ac, bc$

From linear combinations we do not obtain pure effects but only three main effects A, B, C confused with three interactions AB, AC, BC (called *alias pairs*) and we totally lose the third interaction ABC.

Conclusion:

If interactions are negligible, we get the three estimates of the main effects with only 4 tests.

Fractional factorial design can be of great help if we consider that the real world is multi-factored, so that we can obtain more information from a fractional factorial design such as $2^{4-1} = 8$, to study 7 not clean factors (and interactions), totally losing one interaction, than studying 3 factors completely from a full factorial design such as $2^3 = 8$.

In any case, after 5 variables the fractional factorial design is practically inevitable in order to limit the experimentation cost.

6.5.1 Latin Squares Design

A way to use fractional factorial analysis when the factors are, e.g. three and the number of levels is the same for all factors, is the Latin Squares design. We can consider this possibility.

A Latin square is an nn array filled with n different symbols, each occurring exactly once in each row and once in each column[3] [2]. If each entry is written as a triple (A, B, C), where A is the row, B is the column, and C is the symbol, we obtain a set of n^2 triples called the orthogonal array representation of the square.

The Latin square experimental design enables simultaneous analysis of 3 factors at three levels with only $3^2 = 9$ observations. This advantage, represented by saving of material and therefore time and money, is associated with the disadvantage of rigidity of the experiment: all three factors must have the same number of levels.

E.g. a Latin square for 3 factors at 3 levels with 9 treatments is shown in Table 6.9. The corresponding complete factorial design should have $3^3 = 27$ combinations. The present incomplete design, (but square with the same number of levels), has only 9 runs (A_1, B_1, C_1) (A_1, B_2, C_2) (A_1, B_3, C_3) and similarly for A_2 and A_3.

No-replications and no interactions are limits of the present scheme, but the experiment is balanced, since factors and levels are combined the same number of times with the other factors.

In order to have unbiased evaluation when we apply a Latin Square to perform an experiment, we use a *randomization* procedure for assigning treatments, even

[3]The symbols may be letters, numbers, colours, etc. R. A. Fisher promoted the use of Latin squares in experiments in his 1935 book The Design of Experiments. A stained glass window in Gonville and Caius College in Cambridge (UK) commemorates the very important results of his research.

Table 6.9 Latin square 3^2 with three factors A, B, C

	B_1	B_2	B_3
A_1	C_1 y_{111}	C_2 y_{122}	C_3 y_{133}
A_2	C_2 y_{212}	C_3 y_{223}	C_1 y_{231}
A_3	C_3 y_{313}	C_1 y_{321}	C_2 y_{332}

maintaining the structure of the Latin Square. The ideal randomization would be to randomly permute the columns, and randomly permute the rows.

6.5.1.1 Statistical Model for an Unreplicated Latin Square

$$y_{ijk} = \mu + \rho_i + \tau_j + \gamma_k + e_{ijk} \tag{6.20}$$

with:
$i = 1, \ldots, m$ levels of factor A
$j = 1, \ldots, m$ levels of factor B
$k = 1, \ldots, m$ levels of factor C
$e_{ijk} = y_{ijk} - \overline{y}_{i..} - \overline{y}_{.j.} - \overline{y}_{..k} + \overline{y}_{...}$ is an Interaction-like term that has the role of an error.

The approximate model to put in Eq. 6.20 is:

$$y_{ijk} = \overline{y}_{...} + (\overline{y}_{i..} - \overline{y}_{...}) + (\overline{y}_{.j.} - \overline{y}_{...}) + (\overline{y}_{..k} - \overline{y}_{...}) + e_{ijk} \tag{6.21}$$

The ANOVA analysis proceeds in the same way seen for the full factorial design.

We can give this interpretation of the Latin square, Table 6.10:

- The internal factor (inside the cells) C is bound to the other two and can not give rise to any possible combinations that, for 3 factors, would be 27. The Latin square experiment allows only 9 combinations.
- Similarly, with 4 factors at 4 levels we would have a 4×4 Latin square with 16 experiments instead of $4^4 = 256$.

Table 6.10 ANOVA for a latin square design with 3 factors at 3 levels with 9 treatments

Source of variation	Sum of squares	*df*	Mean squares
(A) Rows	SSB_A	2	
(B) Columns	SSB_B	2	
(C) Inside the square	SSB_C	2	
Error	SSW	$8-6=2$	
Total	TSS	8	

- In general if $m \times m$ is the design matrix, the degrees of freedom of the sum of total squares is $m^2 - 1$, and the df (degrees of freedom) of every factor is $m - 1$. Then the maximum number of factors that we can introduce in the experiment is $\frac{m^2-1}{m-1} = m + 1$. E.g. for $m = 3$ the max. number of factors is 4 while for $m = 5$ are 6 and so on.
- Interactions are assumed as not existing, then neglected.
- There are not replications, thus no estimates of experimental error. We assume the interaction-like term e_{ijk} as error, i.e. everything that is not an influence of the variables.
- The design matrix is formed by orthogonal vectors. (i.e. vectors with their scalar products equal to zero, since the projection of a vector on any other is null).

6.6 A Case Study on the DOE Application

This section deals with a case study that demonstrates "recursiveness", i.e. the step-by-step implementation of Design of Experiment in a practical application for a process optimization.

The aim is the fatigue life improvement of a mechanical part treated with a thermo-mechanical surface process composed of nitriding and shot-peening.[4]

We want to optimize the fatigue limit, keeping the first nitriding treatment constant and modifying the values of the two variables that control the second (the peening process): the shots sizes and the peening intensity, measured in Almen units.

The experimental campaign is divided into two phases.

- For the first phase, the output is the compression stress at the surface and a DOE has the task of optimizing the values of the two factors to reach the highest value of it. Flat specimens are suitable for this aim, Fig. 6.12.
- For the second phase, the output is the fatigue limit and a DOE has the task of optimizing the choice of shot peening parameters to achieve the highest value.

[4] Shot peening is a cold working processes in which the surface is bombarded with small shots. The resistance benefit is the result of the effect of compression stress distribution and cold working, both due to the shot induced deformation.

A plane specimen is not the most suitable, since shot-peening is particularly active in shielding the peak stresses that arise in the notches of mechanical components, while it has little effect on flat bodies, without geometric discontinuities. This theoretical conjecture must be experimentally validated, by comparing the results of two types of specimens: plane and notched [4, 5, 11].

6.6.1 Residual Stress Distribution

The material is a 32CrMoV13 alloy steel, with the following chemical composition:

- C = 0.32%
- Cr = 3%
- Mo = 1%
- V = 0.20%

The specimens were nitrided and then subjected to shot-peening treatments, and are flat and rectangular of uniform section designed to be subsequently tested on a four bending fatigue resonance machine, Fig. 6.12.

The input factors of the experiment for optimizing the residual stress distribution are:

- *Almen intensity*: the choice of its parameters is limited by the level of surface finish we accept.[5]
- The first purpose of the investigation was to measure the residual stress on plane surfaces arose by both treatments.
- The first operation of nitriding was made with constant parameters.
- The second operation of shot-peening was done with a factorial design at two levels shown in Table on the left side of Fig. 6.10.

The technological cycle was exactly the following:

- Specimens were manufactured by an electro-erosion machine and then, grounded and nitrided.
- Specimens were shot-peened.
- Measurement of total residual stresses due to the added effect of nitriding and shot-peening was performed using a diffractometric method, removing, step after step, very thin layers of material from the specimen surface by etching.

The calculation matrix of the design with two factors at two levels is shown in Table 6.11. The columns headings are:

[5]The maximum intensity of 16 A was chosen and then its value was decreased to 12 A and 8 A. It is necessary to not exceed the Almen 16 A intensity in order to avoid a subsequent grinding of the surface to remove the peaks of excessive roughness, which would nullify the positive effect of the treatment.

Run	x_1	x_2	ϕ_{mm}	I
1	−1	−1	.3	10
2	+1	−1	.6	10
3	−1	+1	.3	17
4	+1	+1	.6	17

(a) *Design Matrix of the Experiment*

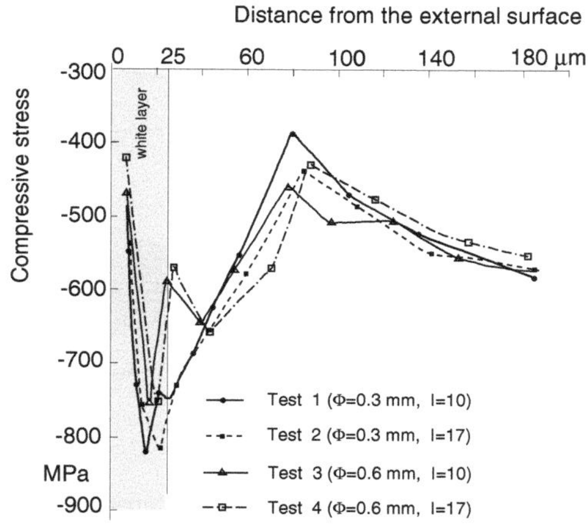

(b) *Distributions of residual stress for shot-peening tests in the Table.*

Fig. 6.10 Design for the analysis of shot-peening treatments. The white layer depth is shown in gray

Table 6.11 Calculation matrix with the measurements of residual stresses and the distance of the stresses from the external surface

T.	x_1	x_2	x_{12}	y_I (MPa)	y_{II} (MPa)	y_{III} (μm)
1	−1	−1	+1	−745	−745	25
2	−1	+1	−1	−754	−754	25
3	+1	−1	−1	−590	−642	38
4	+1	+1	+1	−571	−658	40

- y_I = stress value at 25 μm from the surface, (in MPa).
- y_{II} = maximum compression value (in MPa).
- y_{III} = distance of maximum value below the white-layer (see Fig. 6.10) from external surface (in μm).

The design output is prudentially considered the value of residual stress in points at 25 μm below the surface.[6]

It was not possible to perform measurement replications, due to the difficulty of the measurement method. As a consequence, do not have a "clean" estimate of the experimental error.

[6]This is for taking into account the uncertainty of the residual stress at the surface, due to the presence of the white layer—an amorphous structure generated by the nitriding process— that confuses the answer of the X-ray diffractometer [5].

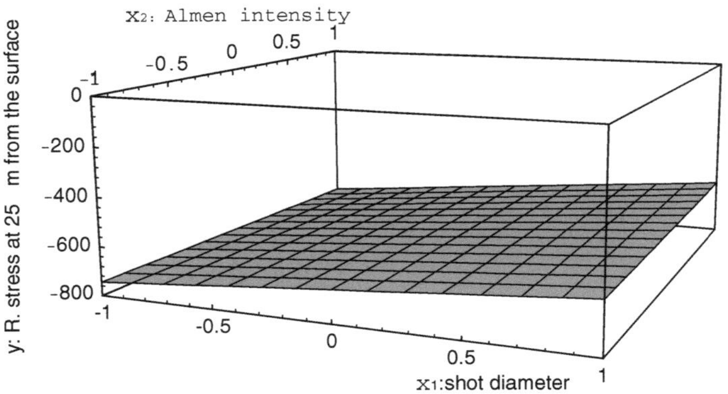

Fig. 6.11 Predicted response surface for stress values at a depth of 25 μm from the surface. Nitrided and shot-peened steel

To overcome this difficulty, an approximate error estimate was made on the basis of an *expert testimony* of similar cases, assuming an error of about 30 MPa.[7] The pooled standard deviation of any individual test was thus assumed equal to $30/6 = 5$ MPa, (assuming an error 6σ), then with a variance $s_p^2 = 25$. Figure 6.11 shows the predicted response surface of the experiment.

For y_I it is described by the following (empirical) mathematical model:

$$y_I = b_0 + b_1\, x_1 + b_2\, x_2 + b_{12}\, x_1 x_2 + \epsilon \tag{6.22}$$

If E_i and E_{ij} are the estimates of main and the interaction effects ($b_i = E_i/2$; $b_{ij} = E_{ij}/2$.) and $b_0 = 1/4(\sum y_i)$, Eq. 6.22 gives, Fig. 6.11:

$$\hat{y}_I = -665 + 84.5\, x_1 + 2.5\, x_2 + 7.0\, x_1 x_2 \tag{6.23}$$

We can see that the shot diameter is the main control parameter of residual stresses together with the interaction effect with the Almen intensity.

6.6.2 Design of Experiment on Fatigue Response of Plane-Surface Specimens

Fatigue limit improvement is the real aim of this research and every thermo-mechanical treatment is justified if it produces a significant increase of fatigue life [11]. Therefore fatigue tests were planned on plain-surface specimens, with the use of a four point bending jig mounted on a testing machine, Fig. 6.12. The same Figure

[7] Hypothetically this value is derived from a large number of observations (∞).

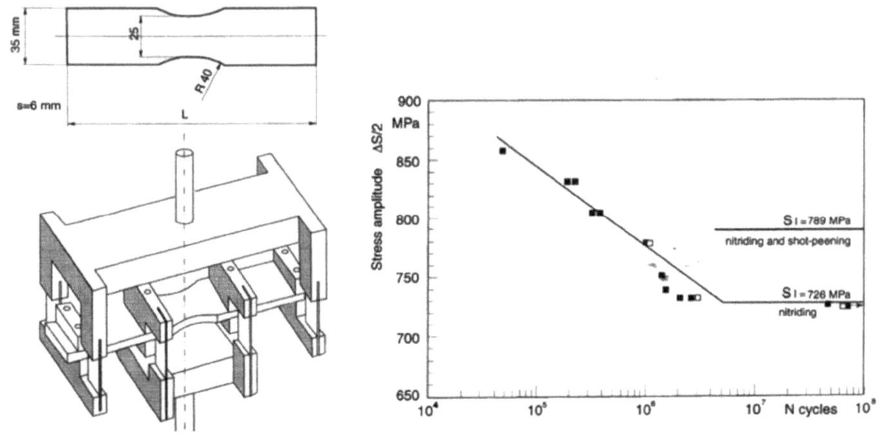

Fig. 6.12 Specimen geometry, bending jig for fatigue tests with a constant plane of bending and fatigue curve for nitrided steel with the fatigue limit before (=726 MPa) and after shot peening (=789 MPa)

Table 6.12 Calculation matrix for fatigue limits

Tests	x_1	x_2	x_{12}	y, fatigue limit
1	−1	−1	+1	764 (772*)
2	−1	+1	−1	799 (808*)
3	+1	−1	−1	801 (837*)
4	+1	+1	+1	791 (770*)

*Values according to Dixon method

shows the complete fatigue curve for the nitrided steel with a comparison between the fatigue limits before and after shot-peening. The increment is relatively low (around 9%). Since there is not full agreement on the method for measuring the fatigue limit, two results are shown in the calculation matrix of this design: The first is obtained by the application of the Hodge-Rosenblatt method [12]. It is recognized by the authors as being reliable and the subsequent calculations are based on the results obtained by this method. Table 6.12 shows the calculation matrix of the full factorial design for two variables at two levels on the Hodge-Rosenblatt results and, between parentheses, a second result obtained by the application of the Dixon method.

The mathematical model for the fatigue limit $\hat{y}$ according to the Hodge-Rosenblatt method, is given by Eq. 6.24.

$$\hat{y} = 788.75 + 7.25\, x_1 + 6.25\, x_2 - 11.25\, x_1 x_2 \tag{6.24}$$

Varying the levels of variables changes the fatigue limits very little. This analysis confirms the previous doubt: both the shot diameter and the Almen intensity, at least within the tested interval, are able to improve the fatigue limit, but a dominant role of one factor does not exist, only the average value of their effect (=788.75) being statistically significant.

6.6.3 *Design of Experiment on Smooth and Notched Specimens*

A different answer must be expected in the case of smooth and notched specimens of different form than the previous one, which was a simple rectangular flat slab where the shot-peening gives a very limited increment of fatigue life.

Therefore, a second Design of experiments was planned to estimate the influence of the same shot-peening variables (shot diameter and Almen intensity) on two different types of specimens: the first one has a uniform cylindrical bar and the second one has a deep notch between two diameters of a cylindrical bar of the same diameter as the first, Fig. 6.13. Both are tested with a constant plane of bending, the same equipment shown before in Fig. 6.12.

The new factorial design 2^2 is developed for 4 points on the domain boundary (white circles) (Tables 6.13, 6.14 and 6.15) plus the gray one in the center as a confirmation test. Figure 6.13b [5]. It has the task of verifying the choice of a design with two factors at two levels.

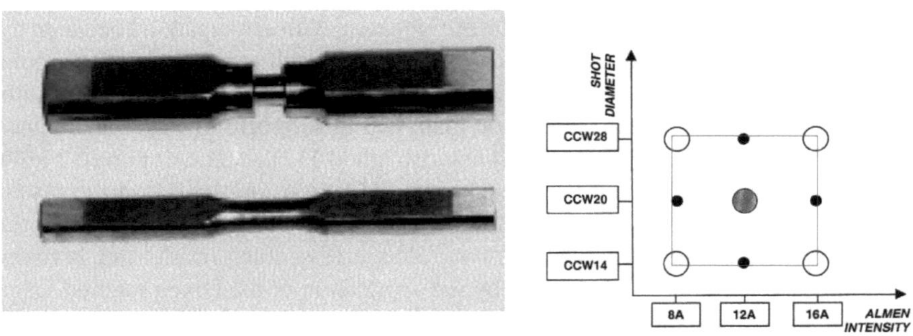

(a) *Shape of notched and smooth specimens. The fatigue tests are performed on the same device shown in Fig. 6.12, with a constant bending moment acting in one plane*

(b) *Design 2^2 (white points) plus a central one as a confirmation test and a full design 3^2 (white and black points) plus the central one*

Fig. 6.13 Smooth and notched specimens with two different design schemes on fatigue response of shot-peening treatments

Table 6.13 Design matrix for unnotched and notched specimens

No.	x_1	x_2	Almen (A)	Shot diam.
1	−	−	8	CW 14
2	+	−	16	CW 14
3	−	+	8	CW 28
4	+	+	16	CW 28

Table 6.14 Calculation matrix for unnotched specimens, without replications

T.C.'s	I	x_1	x_2	x_1x_2	y	$\bar{y}_i$
$i = 1$	+	−	−	+	1	847
2	+	+	−	−	a	823
3	+	−	+	−	b	835
m = 4	+	+	+	+	ab	861
	$\sum/4$	A=$\sum/2$	B=$\sum/2$	AB=$\sum/2$		
	841.5	0.5	6.5	12.5		

The first test was a two factor at two level design with the following design and calculation matrix and the response surface for smooth specimens is:

$$y = 841.4 + 0.5x_1 + 6.5x_2 + 12.5x_1x_2$$

while the response surface for the notched specimen is:

$$y = 785 - 9.5x_1 + 9.5x_2 + 3.5x_1x_2$$

A confirmation test was done in both cases in the center of the factors domain with the following levels: 12A and $CW20$, obtaining the following results:

- 847 for smooth specimen
- 851 for notched specimen

A response ruled-surface was used but the test shows that in the central point of the domain a value of fatigue limit (especially for the notched specimen) is not compatible with this elementary interpolation with a ruled surface. One extra-test in the central point of the domain suggests a more detailed analysis with a 3^2 design.

6.6.3.1 A new Design with Two Factors at Three Levels

This experiment is performed using a factorial 2^3 design with two factors at three levels, with 8 runs, plus a ninth experiment in the central point, with a total number

Table 6.15 Calculation matrix for notched specimens, without replications

No.	I	x_1	x_2	x_1x_2	y	$\bar{y}_i$
$i = 1$	+	−	−	+	1	787
2	+	+	−	−	a	774
3	+	−	+	−	b	793
m = 4	+	+	+	+	ab	787
	$\sum/4$	A=$\sum/2$	B=$\sum/2$	AB=$\sum/2$		
	785	−9.5	9.5	3.5		

Table 6.16 Fatigue limits for smooth and notched specimens after nitriding and shot-peening treatments, with different values of shot-peening parameters

Alem intensity	CW14	CW 20 Shots diam.	CW 28	CW14	CW 20 Shots diam.	CW 28
8 A	847 ± 20	884 ± 22	835 ± 20	787 ± 22	832 ± 17	793 ± 24
12 A	889 ± 24	847 ± 20	884 ± 22	832 ± 17	851 ± 24	822 ± 17
16 A	832 ± 22	854 ± 22	861 ± 24	771 ± 27	852 ± 17	787 ± 26
	Fatigue limit for nitrided smooth specimens: 827±20			Fatigue limit for nitrided notched specimens: 702±20		

of output variables equal to 9 for smooth specimens and 9 for notched specimens. The output of all the tests are shown in Table 6.16.

6.6.3.2 Variance Estimation

Unfortunately we have not done replications of tests (the fatigue tests are very awkward and expensive) and the variance must be estimated in another way. The fatigue limit determination gives the associated variance for every step of limit determination. So it is possible to estimate the variance by weighting the sum of squares of the single errors, related to their degree of freedom.

This case study showed that for notched specimens made of 32CrMoV13 nitrided and then shot-peened steel, an improvement on fatigue life (in addition to the improvement for nitration-only) is achieved by more than 21%, i.e. the controlled shot-peening treatment can appreciably improve the fatigue limit only in notched details where the stress concentration is higher and dangerous for the life of a structural component. For appropriate insights see: [3, 6, 7, 9, 13, 17].

6.7 What We Have Learned from This Chapter

The main concepts on DOE were introduced:

1. *Full Factorial Design*. It is the most complete way to perform a Design of experiments. Nevertheless, when the number of factors is high, a full factorial design requires too large a number of treatment combinations and is not the best choice. This justifies the introduction of:
2. *Fractional Factorial Design, such as Latin Squares Design*. Fractional design is a factorial experiment in which only an adequately chosen fraction of the treatment combinations, required for the complete factorial experiment, is selected to be run. This method allows sufficient information to be obtained on the effects when the interactions are considered negligible.
3. *Response Surface*. It is useful to introduce this concept when variables are continuous. The optimal solution can be outside the experimented points.
4. *Orthogonal arrays*. It is a matrix with orthogonal column vectors [15]. Orthogonality of the vectors of the matrix guarantees non-overlapping, balanced and uncorrelated experiments.
5. The simple *two factors and two levels design* is a useful tool in the early stages of the experiment when the dependence and the number of factors controlling a process are not clear. It allows the minimum number of tests with which the factors can be thoroughly studied. These initial designs are called sorting or identification designs.
6. DOE is very useful to optimize a technological process. The case study shows this type of application.

References

1. 503 S (2018) Design of experiment. https://onlinecourses.science.psu.edu/stat503/node/1
2. Bailey RA (2013) Latin squares. http://www.maths.qmul.ac.uk/~rab/gcs2hand.pdf, g. C. Steward lecture, Gonville and Caius College, Cambridge
3. Berger P (1994) Experimental design: is it important? Lecture handout, MIT Boston Summer Professional Program. http://professional.mit.edu/programs/short-programs/design-and-analysis-experiments
4. Cristofolini L, Croccolo D, Freddi A (2001) Miglioramento della resistenza a fatica di un acciaio niturato e pallinato. una applicazione del progetto dell'esperimento (DOE). Tratt Finit 41(2):81–88
5. Croccolo D, Scazzieri F, Freddi A (1999) Resistenza a fatica per flessione alternata di provini in acciaio 32CrMoV13 nitrurato e pallinato. In: Atti XXVIII Convegno AIAS, Organization AIAS, vol 1
6. DeVor RE, Chang TH, Sutherland JW (1992) Statistical quality design and control. MacMillan Publ. Co, N.Y
7. Dixon W, Massey F (1969) Introduction to statistical analysis. McGraw-Hill, N.Y
8. Finney DJ (1955) Experimental design and its statistical basis. University of Chicago Press, Cambridge
9. Fisher R (1974) The design of experiments, 9th edn. Hafner, N.Y

10. Freddi A (2004) Imparare a progettare, vol 1, 1st edn. Pitagora Editrice Bologna
11. Freddi A, Veschi D, Bandini M, Giovani G (1997) Design of experiment to investigate residual stresses and fatigue life improvement by a surface treatment. Fatigue Fract Eng Mater Struct 20(8):1147–1157
12. Freddi A, Olmi G, Cristofolini L (2015) Experimental stress analysis for materials and structures. In: Stress analysis models for developing design methodologies. Series in solid and structural mechanics, vol 1, 1st edn. Springer
13. Goos P, Jones B (2011) Optimal design of experiments: a case study approach. Wiley, N.Y
14. Hardwick C (2017) Practical design of experiments-Doe made easy, minitab statistical software edn
15. Kacker RN, Lagergren ES, Filliben JJ (1991) Taguchis orthogonal arrays are classical designs of experiments. J Res Natl Inst Stand Technol 96:577591
16. Mason RL, Gunst RF, Hess JL (2003) Statistical design and analysis of experiments: with applications to engineering and science, 2nd edn. Wiley
17. Montgomery DC (1991) Design and analysis of experiments. Wiley
18. Montgomery DC (2012) Design and analysis of experiments (Chapter 12). Wiley
19. Phadke MS (1989) Quality engineering using robust design. PTR Prentice Hall, Enlewood Cliffs, New Jersey
20. Soliani L (2003) Statistica applicata alla ricerca biologica e ambientale. UNI, NOVA Parma

Chapter 7
Introduction to the Taguchi Method

Abstract We now introduce a further powerful engineering tool: the Robustness concept. It is inextricably linked to the name of Genichi Taguchi, a Japanese engineer. Although his approach is based on what was outlined in the previous chapter on Design of Experiment, it contains strong innovation of those concepts. Significant improvements are proposed for three design phases: system design, parameters design and tolerance design. It is not possible to draw a comprehensive picture of the method here. Only the main concepts are introduced, leaving any more complex treatment of the method to specialized books. This method is a real improvement of the application of statistical methods to design applications. This chapter describes the basic concepts, among which the Quality Loss Function that better than other interpretations clarifies the term, often ambiguous, of quality. In this way, the link between robustness, reliability and design of quality becomes obvious. We tried to give emphasis to ideas more than to technicality, but a limited mathematical formalism is necessary.

7.1 Integration of the Taguchi Method with Other Methods

Genichi Taguchi applied the statistical methods and the design of experiments concepts, (that he used in communication theory and practiced in department of a Japanese company and in laboratory of electrical communications and telegraphs), to the product and process design.

The difficulty of his original text slowed the spread of his approach. His book was first translated into Chinese and later to English. Since 1980 the method has been applied only in the United States and later in Europe. Nevertheless, the method captured the attention of large companies that heavily invested in it. Taguchi did not limited the presentation of one point of view: he suggested the use of the entire set of statistical planning tools (introduced before), particularly suited to improve quality of engineering design, essentially for an off-line experimentation.

The applications of the method came first for improving a manufacturing process, although it may in principle be applied to a variety of products and process designs, in prototyping as well in inspection phase [13].

A. Freddi and M. Salmon, *Design Principles and Methodologies*, Springer Tracts in Mechanical Engineering, https://doi.org/10.1007/978-3-319-95342-7_7

The founding concept is the achievement of a minimum sensitivity design, able to control the output variables in the face of random variability of the input.

- The method is integrated with other modern concurrent engineering methods. It also requires, like other methods, a preliminary clarification phase to identify the factors that may affect the system response [9].
- In QFD, the application of this method is suitable for predicting the influence of some factors on results. In interrelation matrix (the roof of the house of quality) it may clarify whether the modification of a variable value produces a positive effect or not.
- In FMEA, once the causes are identified, this approach evaluates the role of not controlled or not controllable variables.
- Compared to the general experimental design methods, the Taguchi approach has the following features and merits [11, 14, 16]:
  - It uses factorial as well as fractional experimental design plans.
  - It does not require the response surface representation nor a mathematical expression of the output.
  - It introduces a new realistic concept of quality with the a definition of Quality Loss Function.
  - It controls the role of interactions between the main factors and noise factors.
  - It classifies the different role of the parameters to reach a low sensitivity.
  - It extends the concept of signal to noise ratio as quality index, from communication to design theory.

The method distinguishes three phases in the design process:

1. System design
2. Parameters design
3. Tolerances design

The *system design* substantially corresponds to conceptual design, i. e. the phase of conceptual variants selection. Up this point, nothing is really new with respect to what has been developed in classic texts. The design up to the prototype establishes the system layout and its controlling parameters. Conceptual design covers innovative use of new technologies, selection of materials and choice of manufacturing processes.

The most original part of the Taguchi method is *parameters design*, also called robust design.

Parameters design is the selection of the control factors and the determination of optimal levels of everything. It is the search for an ideal combination of factors that make the system achieve the standard foreseen for quality and less sensitive (more stable) to disturbance factors, represented by the effects of environmental or non-controllable variables, and of the factors not taken into account as control parameters.

A third phase, *tolerances design* develops further reduction of the system variance when the measures adopted in previous points are insufficient. In this phase it is decided which factors should have tighter tolerance and which should have greater tolerance.[1] Therefore, Robust Design is more than the design of experiment because it reaches two goals:

- It reduces the variability of product (service) functionality with reference to the environment.
- It makes sure that the choices adopted in the laboratory remain the same in production and use [3, 8, 9].

7.2 Quality Loss Function

We have seen that the key concept of the Taguchi method is the reduction of variability on the output variables.

He relates the variability of the actual response to a desired value. There is a *Quality Loss* when quality characteristics (quality indexes) of a product differ from the nominal, either in the event that indexes are greater or lower than the required level. The concept of Taguchi's quality loss function is in contrast with the classic concept of quality:

if a product feature doesn't meet the designed specifications the product is termed of poor quality (rejected), irrespective of deviation entity of its indexes from the target values (mean values of tolerance zone).

If $L(y)$ is the Quality Loss Function, the traditional definition is the following, (goalpost philosophy):

- $L(y) = 0$ within the range
- $L(y) = L_{max}$ outside the range

This view is represented by the step-like function of Fig. 7.1a. This conventional view is inadequate, because it does not consider any quality loss inside the field of tolerance. It is insignificant that this deviation is in favour of the user or of the producer. In both cases it would still represent a loss for the other subject.

For these reasons, it is more appropriate to express such loss with a continuous and symmetrical function, such as:

$$L(y) = k(y - m)^2 \qquad (7.1)$$

where m is the *target value* or *mean value*, y is the actual value of the product and k is a constant, Fig. 7.1. With the Taguchi philosophy:

- Products and services must be designed in such a way that they are robust with respect to variability sources.

---

[1] Obviously, tighter tolerances require additional costs, which in Japanese philosophy are precisely relegated to the role of extreme necessity and should be avoided whenever this is possible. Tolerance is to be understood in the general sense: even the introduction of more valuable materials, more expensive machines and equipment must be interpreted as a tolerance reduction [15].

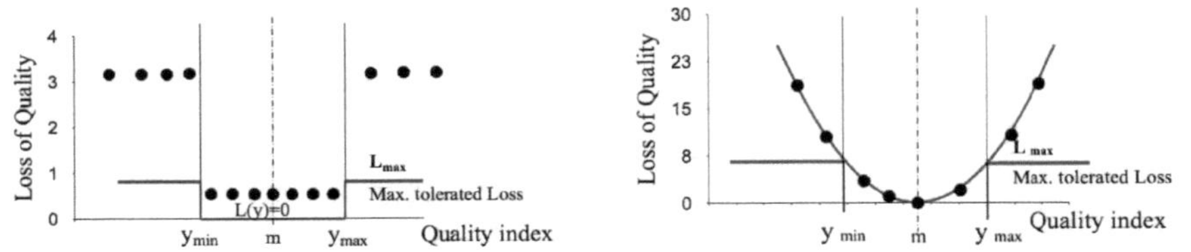

(a) *The conventional view on quality loss: all the results inside the interval* $y_{max} - y_{min}$ *are considered of the same quality*

(b) *The Taguchi view on quality loss: inside the interval* $y_{max} - y_{min}$ *quality is NOT constant*

Fig. 7.1 Two different views of quality

- Experiment-based methods are the most suitable tool to achieve this goal.
- Working with reduced variability is more important than the compliance with the values set in the specification.

The meaning of symbols is the following:

- $L(y)$ the quality loss function
- y quality level
- m target value
- k a scale constant to transform the square difference into loss units.

The quadratic expression describes the continuity of the loss that is zero only at the target value m, and assumes the value L_{max} at the extremes of the tolerance field.

Factor k is a scale factor, determined by the conditions:
for $y = y_{max}$ is $L(y) = L_{max}$ then:

$$k(y_{max} - m)^2 = L_{max}$$

and:

$$k = L_{max}/t^2$$

t being the positive tolerance, equal to the negative one, $t = (y_{max} - m) = (m - y_{min})$. Then:

$$L(y) = \frac{L_{max}}{t^2}(y - m)^2 \tag{7.2}$$

This equation is true for a single product. In the following we will see how we can modify the Eq. 7.1 when the “loss” is to be calculated for multiple products.

7.2.1 *Quality Loss Functions in Three Common Cases*

Three conditions must be distinguished, depending on quality indexes, respectively nominal, minimum or maximum:

1. The best value is the nominal when we want to reach a pre-set nominal value. This is the case previously discussed when the quality loss function was first introduced, but there are other cases of great technical importance:
2. The best value is the minimum (smaller is better): it happens when the target is a negative quality such as noise, consumption, pollution, impurity, etc.
3. The best value is the maximum (larger is better) that is the case of targets such as duration, speed, or others, without defined upper limits.

We want to look at the details of these cases.

7.2.1.1 The Best Value Is the Nominal

In this case, the expression of quality loss function is the previous one, Eq. 7.2 and is zero when the *Quality Index for each single unit* coincides with the desired nominal (target) value.

In the technical literature [8, 15] we can find an extension of the analysis of a single case: in addition to loss analysis for one unit, we want to estimate the expected loss for multiple units and the mean loss for a continuous process. The expression for a unit is replaced by the following, which refers to multiple units:

Multiple Units

$$L(y) = k(MSD) \tag{7.3}$$

where MSD is the *mean squared deviation* of the different units with respect to the nominal value m. MSD for a number n of units is:

$$\begin{aligned} MSD &= \frac{1}{n} \cdot [(y_1 - m)^2 + (y_2 - m)^2 + (y_3 - m)^2 + \cdots + (y_n - m)^2] \\ &= \frac{1}{n} \sum_{i}^{n} (y_i - m)^2 \end{aligned} \tag{7.4}$$

Mean Loss of a Process

For a continuous process with a probability density function $f(y)$ for the quality index y (that is a random value) and a variance, the mean of the process is [8]:

$$L(y) = E[L(Y)] = \int L(y) f(y) dy \tag{7.5}$$

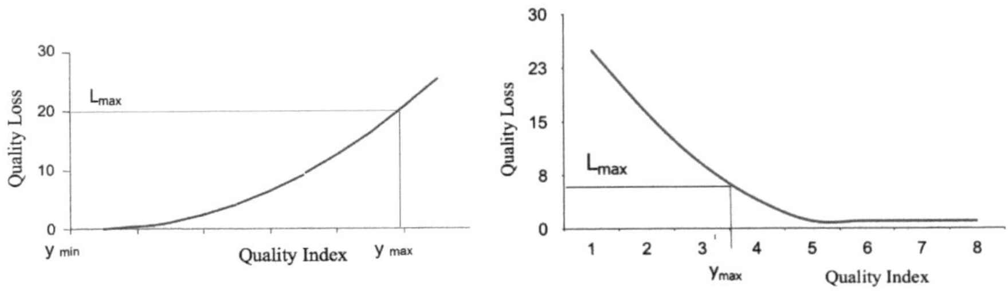

Fig. 7.2 The quality loss function for different cases: smaller is better (on the left) and larger is better (on the right)

If $f(y)$ is normal, the mean loss is equal to the square mean of the difference between y and m (nominal or target) which, in turn, can be subdivided into the sum of the square of the difference between the mean of the output μ and the nominal value m plus the variance of the process, (i.e. the variance of y with respect to the mean μ, indicated with σ^2) [2]:

$$L(y) = k[(\mu - m)^2 + \sigma^2] = k(MSD) \tag{7.6}$$

7.2.1.2 The Best Value Is the Minimum

The case is achieved when the quality index reaches the minimum (*for each single unit*). In this circumstance, the loss function assumes the expression, Fig. 7.2:

$$L(y) = ky^2 \tag{7.7}$$

with: $k = L_{max}/y_{max}^2$

For more units and for a continuous process we have respectively:

$$L(y) = k(MSD) \tag{7.8}$$

where:

$$MSD = (1/n) \cdot (y_1^2 + y_2^2 + y_3^2 + \cdots + y_n^2) = (1/n) \cdot \sum_{i=1}^{n} y_i^2$$

and in the case of continuous distribution ($m = 0$):

$$MSD = \mu^2 + \sigma^2 \tag{7.9}$$

7.2.1.3 The Best Value Is the Maximum

In this case, the loss function assumes the expression, Fig. 7.2:

$$L(y) = k/y^2$$

with: $k = L_{max} \cdot y_{max}^2$
or for a continuous process:

$$L(y) = k(MSD)$$

The mean squared deviation is, respectively:

$$MSD = (1/n) \cdot (1/y_1^2 + 1/y_2^2 + 1/y_n^2) = (1/n) \cdot \sum_{i=1}^{n} (1/y_i^2)$$

and:

$$MSD = 1/\mu^2 \cdot (1 + 3\sigma^2/\mu^2) \tag{7.10}$$

It is clear that, to minimize losses, we must minimize the Mean Square Deviation of the process *MSD*. That means that in the case of the *nominal value is the best* we can consider the distance of the mean of y from the nominal target value m. In more general cases, the closer to the target μ and the variance σ^2 to 0, the less is the loss.

Two main objectives are achieved by the Robust Parameter Design (RPD) approach:

- The mean or average location of the response μ is at the desired level m, and the variation or dispersion of the response σ is as small as possible.
- The consequent strategy is to find settings for the controllable variable such that the variation of response due to noise factors is minimized.

7.3 Robust Parameter Design

According to a common definition, robustness is a characteristic describing a model's, test's or system's ability to effectively perform while its variables or assumptions are altered, so a robust concept can operate without failure under a variety of conditions. When the concept is translated into a system, it refers to the *ability of tolerating perturbations (variations)* that might affect the system's functional body. From the previous analysis, see (7.3), (7.9) and (7.10), it is clear that the critical parameter to minimize losses and increase quality is linked to the ratio of the difference between the desired and the obtained value of the quality index (i.e. to μ value and to variability σ of quality index).

Consequently, Taguchi suggests, as the critical index, *the signal to noise ratio* S/N, defined (in general terms) as the decibel measurement of the ratio between signal power and noise power:

$$S/N = 10\ \log_{10} \frac{Signal\ power}{Noise\ power} \tag{7.11}$$

Signal power and Noise power are generally interpreted as mean value and standard deviation of the index.

The meaning of the relationship is derived via analogy from the theory of communication: the parameter values for the best design are not the ones that maximize the signal but rather the ones that are effective in reducing the noise produced by the variability of the control factors and by variability of not controlled variables of the unchecked entry.

Therefore, the quality of a system is not only given by the mean values of the output variables, but also by the system's ability to keep output variability within acceptable limits without modifying its behavior when the component properties internally (factors) or externally (environmental conditions) vary, as possible effects of different operating conditions. The value of the S/N ratio must be as high as possible, regardless of the types seen of the Quality Loss.

In engineering design possible noise sources can be due to:

- Manufacturing unit-to-unit variations.
- Aging phenomena such as corrosion, etc.
- Fatigue.
- UV degradation.
- Wear.
- Temperature.
- Humidity.
- Environment related as also human interface.

In order to analyze the influence of the noise and to make the process insensitive to the noise, it is better to differentiate:

1. Parameters that remain unchanged in the input of the process (factors which do not affect the S/N ratio or process mean).
2. Parameters that vary but which do not influence the S/N ratio or process mean.
3. Parameters that we want to modify in order to evaluate their effects, called control factors which affect process variability.

7.3.1 Relationship Between S/N and Loss Function

A first question arises:

Does the *maximum* of the signal to noise ratio derive from the *minimum* of the quality loss function? [7, 9, 12, 15].

The answer is no. Minimizing the mean square deviation not necessarily brings μ to coincide with m, because the deviation and mean are often such that decreasing the squared deviation leads to a decrease of the mean, (which may deviate from the desired value).

The signal to noise ratio can be expressed by a general relationship to ensure congruence between the minimum loss and the maximum S/N ratio:

$$S/N = 10\log_{10}(MSD)$$

but in the case of *Nominal is the best*, a discrepancy can be observed: maximizing the noise to signal ratio, i.e. the ratio between the mean value and the squared deviation, does not necessarily imply the minimization of *MSD*.

It is therefore necessary to consider the three cases separately:

1. Nominal is best
   - From the definition of S/N:
     - For one ore more units:

$$S/N = 10\ \log_{10}\frac{\overline{y}^2}{s^2} \tag{7.12}$$

where: $\overline{y} = \frac{1}{n}\sum_{i=1}^{n} y_i \qquad s^2 = \frac{1}{n-1}\sum_{i}^{n}(y_i - \overline{y})^2$

     - For a continuous function:

$$S/N = 10\ \log_{10}\frac{\mu^2}{\sigma^2} \tag{7.13}$$

   - from the definition of *MSD*:

$$S/N = -10\log_{10}(MSD) = \tag{7.14}$$

     - For one or more units:

$$S/N = -10\ \log_{10}\frac{1}{n}\sum_{i=1}^{n}(y_i - m)^2 \tag{7.15}$$

     - For a continuous function:

$$S/N = -10 \log_{10}[(\mu - m)^2 + \sigma^2] \quad (7.16)$$

2. Smaller is better
   - For one or more units:

$$S/N = -10 \log_{10} \frac{1}{n} \sum_{i=1}^{n} y_i^2 \quad (7.17)$$

   - For a continuous function:

$$S/N = -10 \log_{10}[\mu^2 + \sigma^2] \quad (7.18)$$

3. Larger is better
   - For one or more units:

$$S/N = -10 \log_{10} \frac{1}{n} \sum_{i=1}^{n} 1/y_i^2 \quad (7.19)$$

   - For a continuous function:

$$S/N = -10 \log_{10} \frac{1}{\mu^2}(1 + 3^2/\mu^2) \quad (7.20)$$

7.3.2 Optimization Strategy

Control Factors can be divided into four classes:

1. Factors affecting only the mean but not the standard deviation, (Leverage factors).
2. Factors affecting only the standard deviation.
3. Factors affecting both the mean and the standard deviation:
   - Proportionally (Scale factors).
   - Non-proportionally.
4. Factors affecting neither the mean nor the standard deviation (Ineffective levers: as their value changes, performance does not change).

Note: The latter, which are not useful at this stage, can serve to optimize other aspects such as costs. For example, increasing the tolerances on the factors that reduce the costs, but not other factors. They are free parameters that we can choose without affecting performance.

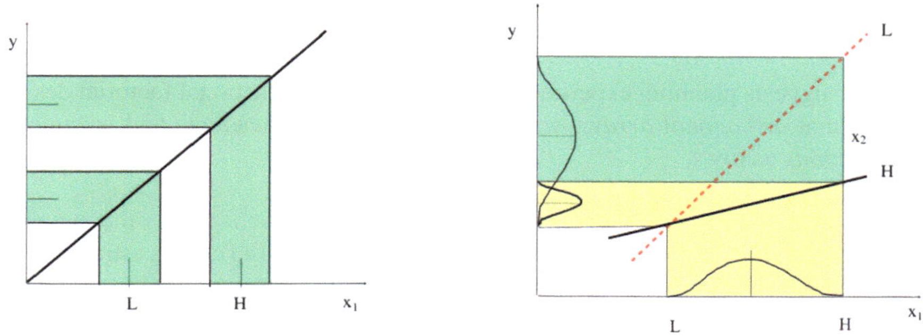

(a) *Factors affecting only Means (Leverage factors)*

(b) *Factors affecting only Standard Deviations*

Fig. 7.3 First types of scale factors. Modified from [4]

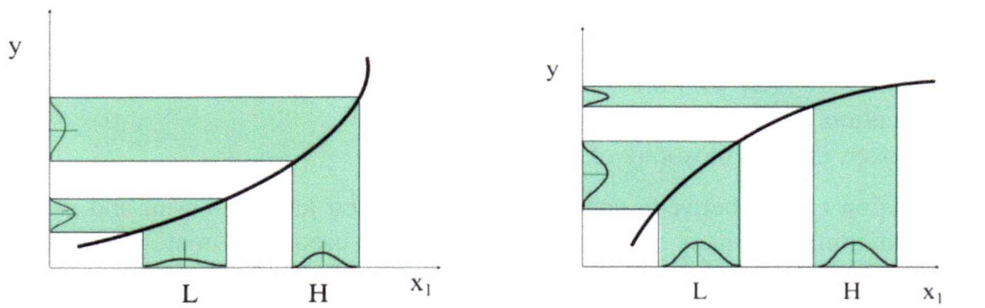

(a) *Factors affecting proportionally St.Dev.'s and Means*

(b) *Factors affecting non proportionally St.Dev.'s and Means*

Fig. 7.4 Second types of scale factors: both st. dev.'s and means affected by the non-linearity of the response. Modified from [4]

The previous discussion on parameter types suggests an optimization strategy.

First, we need to look at the control factors that affect the S/N ratio and the values of these factors that maximize S/N;

Then, the leverage factors must be used (Fig. 7.3) affecting the mean but not the S/N ratio and the relative values of those factors that bring the mean value to coincide with the nominal (target).

If Standard Deviation and Mean do not vary proportionally, Fig. 7.4b, it is convenient to use factor x_1 at level H, rather than at level L, to have a small variation on the output, even with a wide variation on input. The non-linearity of the relationship between input and output allows the reduction of output variability even with great variability of the input.

7.4 Orthogonal Arrays

Taguchi suggests planning experiments by using a highly fractional factorial design with special *orthogonal arrays*, with *factors* as columns headings and *treatments combinations* as rows.

A note on symbols: without risk of confusion, in Taguchi arrays the same symbol (A, B, C etc.) is adopted for representing *factors* and effects estimates. The column headings will be indicated, from this paragraph onwards, by simple sequence of numbers $(1, 2, 3, \ldots, n)$.

Orthogonal arrays are arrays with orthogonal columns vectors. They have this property: the scalar product of each pair of vectors is equal to zero. That means that the projection of one vector on the other is null.[2] Each vector is then linearly independent and conveys information without redundancy. The so-called Taguchi arrays L that have permutations of the factor levels so chosen that their responses are uncorrelated are built with the same property: each treatment gives a unique piece of information.

Moreover, in the philosophy of Taguchi:

- Interaction effects between control factors are often less important and can be ignored, almost in many cases. So a design can be used to estimate main effects using only a few experimental runs.
- The schemes of this type are not only applicable to two level factorial designs; they can also investigate the main effects when factors have more than two levels.
- The meaning of the symbols of L arrays is the following: *the subscript represents the row number* (i.e. number of t.c.'s). *The exponent is the number of columns.* (E.g. matrix $L_8(2^7)$ in Table 7.3).
  - In the *L*4 array shown in Fig. 7.5a is denoted as $L_4(2^3)$. *L*4 means the design requires four runs. 2^3 indicates that the design estimates up to three main effects at two levels each. It can be used to estimate three main effects using four runs, provided that the two factors and three factors interactions are ignored.
  - Figure 7.5b shows a design that requires twelve runs and can be used to estimate up to eleven effects [10].
  - The number of columns is the number of rows minus one, (e.g. $2^3 - 1 = 7$) then is the number of *factors* as well as the number of *estimates* that can be made by means of one experiment. So the design is simply a matter of picking which factors head which columns.

[2] $\sum_{i=1}^{n} x_{ik} \cdot x_{il} = 0$.

Run	Columns		
	1	2	3
1	1	1	1
2	1	2	2
3	2	1	2
4	2	2	1

(a) *Taguchi Array with $L_4(2^3)$ for 4 runs and 3 Column Vectors*

Run	Columns										
	1	2	3	4	5	6	7	8	9	10	11
1	1	1	1	1	1	1	1	1	1	1	1
2	1	1	1	1	1	2	2	2	2	2	2
3	1	1	2	2	2	1	1	1	2	2	2
4	1	2	1	2	2	1	2	2	1	1	2
5	1	2	2	1	2	2	1	2	1	2	1
6	1	2	2	2	1	2	2	1	2	1	1
7	2	1	2	2	1	1	2	2	1	2	1
8	2	1	2	1	2	2	2	1	1	1	2
9	2	1	1	2	2	2	1	2	2	1	1
10	2	2	2	1	1	1	1	2	2	1	2
11	2	2	1	2	1	2	1	1	1	2	2
12	2	2	1	1	2	1	2	1	2	2	1

(b) *Taguchi Array with $L_{12}(2^{11})$ for 12 runs and 11 Column Vectors*

Fig. 7.5 Two examples of Taguchi arrays

Some of the orthogonal arrays suggested by Taguchi are the following: Fig. 7.5 [10]

- For two levels[3]:
  - $L_4(2^3)$, with 2^{3-1} runs.
  - $L_8(2^7)$, with 2^{7-4} runs.
  - $L_{16}(2^{15})$, with 2^{15-11} runs.
  - $L_{12}(2^{11})$, with 12 runs.
- For 3 levels:
  - $L_9(3^4)$, with $3^{4-2} = 9$ runs.
  - $L_{27}(3^{13})$, with $3^{13-9} = 81$ runs.

The orthogonality of these Taguchi arrays is clear if you reduce them to the form of $-1, 0, +1$ with a subtraction from the all terms of the matrix of a number (2). E.g. the $L_9(3^4)$ array can be reduced to the form shown in Table 7.1 with a vector translation of -2. In this representation it is easy to verify that the scalar products of each pair of column vectors is zero. Taguchi arrays derive from fractional factorial designs, even with some modifications and permutations of columns [10].

[3]The last one when all interactions are assumed to be zero.

Table 7.1 $L_9(3^4)$, matrix with 3^{4-2} fractional treatments in Taguchi form and after subtraction from the all terms of the matrix (-2) of -2 to show the orthogonality property of the vectors

#	1	2	3	4	#	1	2	3	4
1	1	1	1	1	1	−	−	−	−
2	1	2	2	2	2	−	0	0	0
3	1	3	3	3	3	−	+	+	+
4	2	1	2	3	4	0	−	0	+
5	2	2	3	1	5	0	0	+	−
6	2	3	1	2	6	0	+	−	0
7	3	1	3	2	7	+	−	+	0
8	3	2	1	3	8	+	0	−	+
9	3	3	2	1	9	+	+	0	−

Table 7.2 $L_9(3^4)$, matrix with 3^{4-2} reduced treatments, without interactions

#	A	B	C	D
1	1	1	1	1
2	1	2	2	2
3	1	3	3	3
4	2	1	2	3
5	2	2	3	1
6	2	3	1	2
7	3	1	3	2
8	3	2	1	3
9	3	3	2	1

7.4.1 Design with No-Interactions

The *L9* array is shown in in Table 7.2. It can be used to estimate four main effects using nine runs, provided that the two factor and three factor interactions can be ignored.

In the first example:

# of runs: $3^{4-2} = 9$

# of estimates 4.[4]

Another example of orthogonal array suggested by Taguchi is the following $L_8(2^7)$, Table 7.3.

This is a design with an array 2^{7-4}

Variables: $8 - 1 = 7$

[4]Factors can be arbitrarily assigned to columns, (remember that, without confusion, factors are indicated with the same letters as the effects).

Table 7.3 Design with $L_8(2^7)$ orthogonal array

#	1	2	3	4	5	6	7
1	−	−	−	−	−	−	−
2	−	−	−	+	+	+	+
3	−	+	+	−	−	+	+
4	−	+	+	+	+	−	−
5	+	−	+	−	+	−	+
6	+	−	+	+	−	+	−
7	+	+	−	−	+	+	−
8	+	+	−	+	−	−	+

Table 7.4 Design $L_8(2^7)$ without interactions

#	A	B	C	D	E	F	G
	1	2	3	4	5	6	7
1	1	1	1	1	1	1	1
2	1	1	1	2	2	2	2
3	1	2	2	1	1	2	2
4	1	2	2	2	2	1	1
5	2	1	2	1	2	1	2
6	2	1	2	2	1	2	1
7	2	2	1	1	2	2	1
8	2	2	1	2	1	1	2

# of experiments: 8
# of estimates $8 - 1 = 7$

The L_8 array can be used to estimate seven main effects using eight runs, provided that the two factors and higher factors interactions can be ignored. For a design with seven factors/effects: (A, B, C, D, E, F, G) the matrix is shown in Table 7.4.[5]

7.4.2 Design with Interactions

In this case we need a method to decide in which column head the effects must be entered.

Taguchi provided a series of *linear graphs* to orient the effects distribution in every *orthogonal array*, Fig. 7.6. # Main factors: A, B, C, D
Interaction effects: AB, AC, BC
Total t.c.'s: $2^4 = 16$

[5]In the Table, again, letters are used to represent factors and effects.

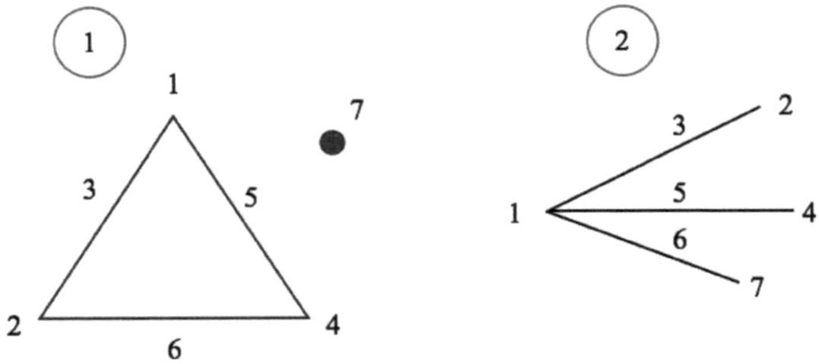

Fig. 7.6 Linear graphs for $L_8(2^7)$

Table 7.5 Design $L_8(2^7)$ with interaction, (graph on the left side of Fig. 7.6)

Exp.	A	B	AB	C	AC	BC	D
No.	1	2	3	4	5	6	7
1	1	1	1	1	1	1	1
2	1	1	1	2	2	2	2
3	1	2	2	1	1	2	2
4	1	2	2	2	2	1	1
5	2	1	2	1	2	1	2
6	2	1	2	2	1	2	1
7	2	2	1	1	2	2	1
8	2	2	1	2	1	1	2

Total Effects: $2^4 - 1 = 15$
Lost effect $1 := ABCD$
Surviving effects: $15 - 1 = 14$ (of these 3 are zero ($CD = BD = AD = 0$)
3 interactions of third order are zero: ($ABC = ABD = BCD = 0$)
Estimations: A, B, C, D, AB, AC, BC
$AB = CD = 0$: we must read this expression in this way: since interaction AB is in pair (called aliased) with CD that is zero, then AB is clean.
$AC = BD = 0$ since BD is zero then AC is clean.
$BC = AD = 0$ since $AD = 0$ then BC is clean.

We adopt the scheme suggested by the first graph: at the vertexes of the graph the factors with no zero interactions and in point 7 the factor D. The design is shown in Table 7.5.

7.5 Robustness Index

Given a system (a process or a product), the signal-to-noise ratio is taken as the most significant parameter to evaluate its robustness, even if it is a good practice to evaluate the mean and variability of the response separately. From a general point of view, for the mean we search for a "good" value (close as possible to a specified target), while for the variability we aim for a minimum value. For multiple units from Eqs. 7.12 to 7.19, we refer to the means values y_i and to the variances s_i of *each experiment*.

We distinguish:

1. *Controllable factors*, whose values can be set before the experiments as well as in the field when the system is operating.
2. *Noise factors* that can be changed during the experiments, e.g. in the laboratory or in a preliminary tuning phase, but not during actual operation of the systems in the field, where they have random variations inside specified limits. Typical noise factors are environmental factors such as temperature or humidity.
3. Variability on the main factors, *tolerances or errors in manufacturing phase* of a product can also be considered noise factors which, like external factors, can influence the product performance.

This is the rationale that justifies the designers' special interest in interactions among controllable factors and noise. Robustness is, in fact, the *ability of tolerating perturbations (noise)*.

This analysis will be developed through an example of Crossed Array Design.

7.5.1 *Crossed Design*

A scheme that allows the analysis of these interactions is the so-called crossed array design.

In a crossed array design, an inner array contains the controllable factors while an outer array contains the noise factors [1] These two arrays are *crossed* because every treatment combination for the controllable factors are run for every treatment combination for the noise factors. Table 7.6 is an example of crossed array design, where the inner array consists of four controllable factors and outer array three noise factors.

An alternative to this design is *a combined array design*: these forms are not discussed in this primer approach [5, 6]. Two examples of pictorial representations of a Taguchi experiment are given in Fig. 7.7. In (a) both inner array and external array are 2^2 designs, while in (b) both arrays are 2^3 designs.

Table 7.6 Crossed design by Taguchi on maximization of a Force

Control factors	Levels		
A = Interference	Low	Middle	High
B = Tube thickness	Thin	Middle	Thick
C = Insertion length	Reduced	Middle	Large
D = Percentage of adhesive	Low	Middle	High
Uncontrolled factors	**Levels**		
E = Conditioning time (h)	24	120	
F = Conditioning temperature (°F)	72	150	
G = Relative humidity of conditioning (%)	25	75	

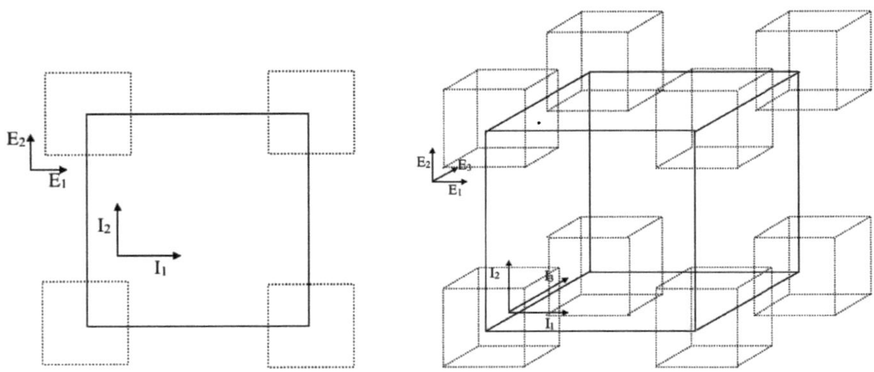

(a) *Crossed Design with Internal Array I*: 2^2 *and External Array E*: 2^2

(b) *Crossed Design with Internal Array I*: 2^3 *and External Array E*: 2^3

Fig. 7.7 Possible arrangements of crossed design

7.5.2 A Taguchi Example of Crossed Design

A case study given by Taguchi concerns the design of an elastomer sleeve mounted by forcing on a nylon tube. The aim of experiment is maximization of pulling force. In this case, we assume the S/N ratio as a parameter to be optimized, in the form given by Eq. 7.19, since the best is the maximum value of extraction force: i.e. the system is optimized when the force is as large as possible and the optimal levels of the parameters are those that maximize the S/N ratio, Table 7.6.

The experimental design was set up through a crossed design, with an inner array $L_9(3^4)$ with four control factors (A, B, C, D) at three levels $(1, 2, 3)$ and an outer

Table 7.7 Internal and external arrays in Taguchi

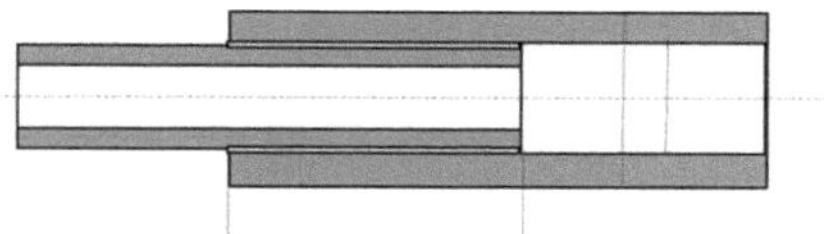

Orthogonal matrix $L_9 (3^4)$					Orthogonal matrix $L_8 (2^7)$							
Combi intern	**A**	**B**	**C**	**D**	**Combi extern al**	**E**	**F**	**EF**	**G**	**EG**	**FG**	**EFG**
1	1	1	1	1	**1**	1	1	1	1	1	1	1
2	1	2	2	2	**2**	1	1	1	2	2	2	2
3	1	3	3	3	**3**	1	2	2	1	1	2	2
4	2	1	2	3	**4**	1	2	2	2	2	1	1
5	2	2	3	1	**5**	2	1	2	1	2	1	2
6	2	3	1	2	**6**	2	1	2	2	1	2	1
7	3	1	3	2	**7**	2	2	1	1	2	2	1
8	3	2	1	3	**8**	2	2	1	2	1	1	2
9	3	3	2	1								

array with three noise factors (E, F, G) at two levels $(1, 2)$. The inner design is a fractional design with only main factors, and the outer one is a full factorial design, Table 7.7.

The internal matrix does not allow interaction determination because all interaction effects are in pairs (aliased) with the main effects (A, B, C, D). It can be accepted when the interactions between the main factors are supposed to be zero.

Vice-versa, the external matrix of the noise factors (2^3), (on the right side of the same Table 7.7) is a full factorial design with the following clean main effects and interactions: E, F, G, EF, EG, FG, EFG.

The experimental results of the crossed design are in 72 cells of Table 7.8. The last two columns show the final output of the experiment, i.e. the mean $\bar{y}$ and the S/N ratio of every row. Among them, 22.82 and 26.91 are the maximum values.

Each row of the results array in Table 7.8 gives the values of y in each vertex of the main cube of Fig. 7.7b; e.g. the value 15.6 in the first cell is the value of the extraction force for the lower level of each control factor (A_L, B_L, C_L, D_L) when the noise factors are at their minimum level, and so on for the others.

A similar analysis [15] can be performed among the y values for discovering the so-called marginal means, i.e. the dependency of the extraction force and of the S/N ratio on one factor at a time:

Table 7.8 Crossed design with only main factors

					External matrix L8										
					E	1	1	1	1	2	2	2	2		
					F	1	1	2	2	1	1	2	2		
					G	1	2	1	2	1	2	1	2		
	Internal matrix L9													Output	
Combi	A	B	C	D										$\bar{y}$	S/N
1	1	1	1	1		15.6	9.5	16.9	19.9	19.6	19.6	20.0	19.1	17.52	24.02
2	1	2	2	2		15.0	16.2	19.4	19.2	19.7	19.8	24.2	21.9	19.47	25.52
3	1	3	3	3		16.3	16.7	19.1	15.6	22.6	18.2	23.3	20.4	19.02	25.33
4	2	1	2	3		18.3	17.4	18.9	18.6	21.0	18.9	23.2	24.7	20.12	25.90
5	2	2	3	1		19.7	18.6	19.4	25.1	25.6	21.4	27.5	25.3	22.82	26.91
6	2	3	1	2		16.2	16.3	20.0	19.8	14.7	19.6	22.5	24.7	19.22	25.33
7	3	1	3	2		16.4	19.1	18.4	23.6	16.8	18.6	24.3	21.6	19.85	25.71
8	3	2	1	3		14.2	15.6	15.1	16.8	17.8	19.6	23.2	24.2	18.34	24.85
9	3	3	2	1		16.1	19.9	19.3	17.3	23.1	22.7	22.6	28.6	21.20	26.15

interference (A), sleeve thickness (B), sleeve length (C) and percentage of adhesive (D) between sleeve and tube.

7.5.3 *Confirmation Experiment*

The Taguchi method always presumes at least one confirmation experiment to control the simplification introduced by fractional design and the results given by $\bar{y}$ and S/N.

Experimentation has an iterative and sequential nature. Often a design of an experiment starts with factorial plans with several variables on two levels and, after the first attempts, it can be reduced to a factorial plan with few variables but at multiple levels. It is the progress of experimentation that sheds light on the use of the most effective experimental design.

Note on Combined Array Design

However, it can be seen that crossed array design may result in a large number of runs even for a fairly small number of controllable and noise factors.

An alternative for these designs are combined array designs which are discussed in the specialized book. The combined array design approach treats all the variables the same, no matter whether they are controllable or noise. These models are capable of modeling the main effects of controllable and noise factors and also their interactions [1].

7.5.3.1 Observation on the Variance Estimation in the DOE of the Previous Chapter

Let's assume that the error is the same for every observation (i.e. the true variance is the same for all observations) and that the observations are independent. We can check this assumption of equal variance through a statistical test. In this hypothesis, each of the sample variances calculated is an independent estimation of the common variance, σ_y^2. A more precise, single estimate of this common variance may be obtained by combining the four independent estimates of the variance. The "pooled" sample standard deviation, s_p is given by:

$$s_p = \left(\frac{\nu_1\hat{\sigma}_1^2 + \nu_2\hat{\sigma}_2^2 + \nu_3\hat{\sigma}_3^2 + \nu_4\hat{\sigma}_4^2}{\nu_1 + \nu_2 + \nu_3 + \nu_4}\right)^{1/2}$$

ν are the degrees of freedom.

7.5.3.2 Variance Estimate for Fatigue Limit Determination on Notched Specimens

The degrees of freedom ν_i is assumed 5, since 6 specimens were used for each test ($\nu_i = 6 - 1$). This provided an estimate of the standard deviation of:

$$s_p = 23.8 \text{ MPa}$$

You can have an approximation of signal to noise ratios at different points of the curve, using the value of the general standard variation previously calculated, equal to $s_p = 23.8$ MPa for notched ones. From Table 6.16 of Chap. 6 the corresponding ratios (calculated by means of Eq. 7.12), are shown in Fig. 7.8.

The highest values of the ratios S/N are obtained with the pairs $(12A,\ CW20)$ and $(16A,\ CW20)$.

Fig. 7.8 Signal to Noise Ratio in the nine points of the design of fatigue limits in the specimens notch, (nitrided and shot peened)

	CW 14	CW 20	CW 28
8A	30.4	30.8	30.5
12A	30.8	31.0	30.7
16A	30.2	31.1	30.4

7.6 What We Have Learned from This Chapter

The Taguchi method, although linked to the DOE method, presents numerous innovations that it are interesting to underline. The method fits into the other methods of the so-called concurrent engineering and constitutes a valid support for them.

The Quality Loss Function is the most realistic interpretation of the concept of quality with a considerable influence on modern methods of design. The signal-to-noise ratio simplifies statistical analysis and, even if it sometimes meets the criticism of statisticians, represents a great simplification that brings designers closer to statistics.

The concept of robust design, which makes designers aware of the concept of noise, can be deepened with the design of tolerances and adjustments. It is still a challenge for the future.

References

1. 503 S (2018) Design of experiment. https://onlinecourses.science.psu.edu/stat503/node/1
2. DeVor RE, Chang TH, Sutherland JW (1992) Statistical quality design and control. MacMillann Publ. Co, N.Y.
3. Foster ST Jr (2001) Design of experiments. http://slideplayer.com/slide/2813463/
4. Freddi A (2004) Imparare a progettare, vol 1, 1st edn. Pitagora Editrice Bologna
5. Hicks CR, Turner K Jr (1999) Fundamental concepts in the design of experiments, 5th edn. Oxford University Press, New York
6. Montgomery DC (2012) Design and analysis of experiments (Chapter 12). Wiley
7. Nair VN (1992) Taguchi's parameter design: a panel discussion. Technometrics 34(2)
8. Phadke MS (1989) Quality engineering using robust design. PTR Prentice Hall, Enlewood Cliffs, New Jersey
9. Phadke MS (2017) Introduction to robust design (Taguchi method) iSix sigma. Job Shop. https://www.isixsigma.com/methodology/robust-design-taguchi-method/introduction-robust-design-taguchi-method/
10. ReliaSoft (2017) Experiment design and analysis reference. http://reliawiki.org/index.php/Experiment_Design_and_Analysis_Reference
11. Robertson R (2002) The Taguchi method. https://www.slideshare.net/MentariPagi4/taguchi-method-14619204
12. Ross PJ (1988) Taguchi techniques for quality engineering. Mc Graw Hill
13. Roy R (2017) A primer on Taguchi method. Nutek, Inc. Bloomfield Hills, MI. USA. http://nutek-us.com/wp-txt.html
14. Subbarao PMV (2009) Design of experiments. https://www.scribd.com/presentation/292423891/mel705-14-ppt
15. Taguchi G (1990) Introduction to quality engineering. Asian Productivity Organization, Japan
16. Wyrwa B (2012) Taguchi method. https://www.slideshare.net/bwyrwa/taguchi-presentation

Part III
Case Studies

The following chapters illustrate some case studies. Chapter 8 deals with the design of testing equipment for structural tests on mechanical components.

This example points out the difference between product designs for final consumers and for laboratory use. The solution of the problem is application of inventive ideas, rather than industrial innovation.

Chapters 9 and 10 deal with the design of an instrumental item (packaging machinery).

In all these examples, the key concept is "One Thing Leads to Another" since the new solutions derive from existing solutions, upgraded in simplicity, cost, production efficiency, and operational control. At the time of development of the last example (a cigarettes packer), the computer science applied to automatic machines was at the initial stage but, due to the foresight of designers, it was adopted with new functionality, better usability, and safety. Because the machine was developed under the design responsibility of one of the authors, a great deal of backstage and proprietary information is made available here. Nowadays, all patents have expired and it is possible to explain the rationale behind many design solutions. By permission of the company management, a rich series of documents have become available. In particular, the product handbook supplied to the authors by the company would have never been available for any more recent products because of obvious confidentiality limitations. Today, at the time of the "fourth industrial revolution," certain solutions are well known but in the proposed example they show the seminal origin of an epochal passage.

The last two chapters devoted to this industrial product have the form of a story that, on the one hand, should not get lost in useless descriptions: *The art of being wise is the art of knowing what to overlook*, William James, (1842–1910) but, on the other hand, should give a sufficient description of the design process.

An appendix is then justified to briefly describe the history of the automatic machine industry in a specific geographical region, where company dynamics and strategies originally led to small companies and later to their amalgamation in big international industrial groups.

Chapter 8
One-Off Product Design

Abstract In the seminal conceptual idea that things derive from other things, in this chapter we will show examples of design that have been developed at the level of useful prototypes. They remained at this level, since they were designed as single units and it was considered not suitable at that time to transform them into large batch products. Only later, did some of them become first models of others, but outside the responsibility of the authors. Nevertheless, these prototypes have received confirmation of their usefulness in special environments, (as equipment in national laboratories and private companies' research centers).

8.1 Introduction

Before addressing the items of the chapter, some clarifications are needed.[1]

> In a primer approach, the authors chose to show only the skeleton of the systematic approach to design. According to the technical literature [10] only when design concerns an adaptive or a variant solutions of existing ones and not something that must be totally original and new, can the path of a systematic design be reduced. (Adaptive solution means to adapt the embodiment design to changed requirements and variant solutions when only size and assemblies are varied.) The design of this type of artifacts follows a process similar to that identified in the *evolution of the species*: each new artifact necessarily derived from the previous artifacts, by means of a process of adaptation and survival on the market. This change is not simple to carry out and sometimes is due to real new inventions and new materials. Totally new solutions come from new discoveries of science that open the way to totally new solutions.

[1] *Henri Poincaré*, about Creativity: Creativity is a new relationship between things that already exist. In [3].

A. Freddi and M. Salmon, *Design Principles and Methodologies*, Springer Tracts in Mechanical Engineering, https://doi.org/10.1007/978-3-319-95342-7_8

The proposed method may be judged simplistic, because it seems to limit the invention to modest adaptations of known solutions. In physical sciences, one should say that it moves in the area of a normal science within a known paradigm [6].

We consider this method the most suitable for a learning phase (primer) of designer education for the following reasons:

- The description of a functional structure is much easier for existing designs than for totally new and totally imagined, not yet existing designs.
- A functional structure of a non-existing artifact can easily lead to an abstract combination of principles totally divorced from reality. Creativity is a complex concept that cannot be reduced to "not yet existing things". From abstraction to absurdity (without the support of a strong experience) the step can be short. This way which is suggested by the systematic method, (interpreted in the hardest and purest way), is therefore suitable for very trained and experienced people, and not for being undertaken at that initial level of training.
- An analysis of pros and cons of *existing solutions* is a great training exercise.
- The paradigm shifts occur as a result of great technical revolutions. Just to mention a few: new components that have transformed the technique, new methods (e.g. of energy storage), new actuators based on new principles, digital devices instead of analog ones, widespread computer science, new possibility of data transmission, microelectronics, etc.
- Last but not least, the proposed method does not exclude or damage the learning of subsequent refinements and abstractions.

The development of a design up to prototypes needs phases of analysis and synthesis, solution comparison and selection of the best ones. It recalls a process driven by evolution, due to a number of criteria that determine success or failure of the products, including but not limited to technical, economic and aesthetic ones must be primarily considered. The success of a product depends on a multitude of other factors that will be examined when we will speak of innovation and of the user's reactions that often follow hardly predictable logic. The analysis of this chapter is limited to cases that make use of technical and economic considerations for one-off products, built for laboratory testing at a prototype level.

Ultimately, the success of a new design can be judged only after the presentation on the market, when a product or a system demonstrates its ability to be appreciated and to survive. Only then, can we speak of effectiveness of an innovation.

8.2 Design of One-Off and Small Batch Products

Unlike products to be made in large quantities (large batch or mass production), products for the construction of only one or a few samples do not require an extensive study of economic convenience, but nevertheless, require a full design development of functionality and safety as well as the definition of all embodiment details.

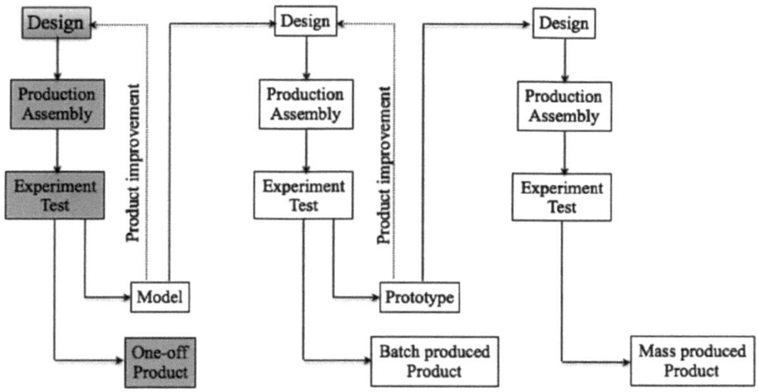

Fig. 8.1 Stepwise development of design of different batch sizes. From [10], ©Springer Verlag

This concept is well clarified by Fig. 8.1 taken from [10] that shows the stepwise developments of design when product is intended for one-off or batch-produced or mass-produced products. Design requires a different number of steps in the three cases.

We limit the analysis to the first branch of the scheme for presenting the case studies of two pieces of equipment. They require the development of the whole design process, from technical specifications, conceptual design and prototype building, but without the industrialization effort that would be necessary for batch or mass produced products.

8.2.1 *Case Study of Testing Equipment Design for Pipe Connectors*

Tubing, piping, couplings and casing connections used for offshore oil exploration and drilling must withstand extremely inhospitable operating environments. In order to assure structural integrity, a number of tests are required [7] The procedures for testing casing and tubing connections is strictly standardized (Technical Committee ISO/TC 67, with Subcommittee SC 5, Casing, tubing and drill pipe, and indication ISO 13679:2002), based on improvements to API Recommended Practice 5C5. They concern materials, equipment and offshore structures for petroleum, petrochemical and natural gas industries.

Testing machines have been developed all over the world for casing and tubing in full scale with combined loads of tension, compression, bending, temperature, internal and external pressure.

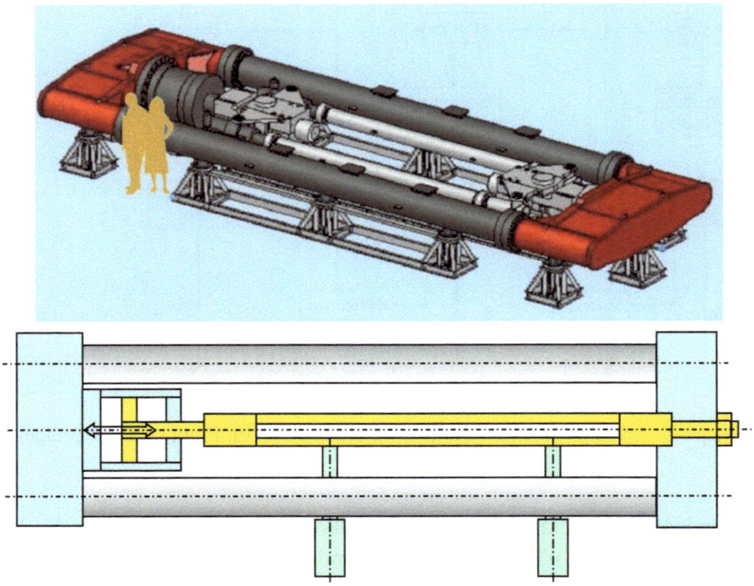

Fig. 8.2 First layout-drawing and sketch- of testing machine for pipes equipped with connection (Giuliani-Forlí-Italy)

This is an example of technical specifications, drawn up directly by the engineering department together with a customer that works closely with it [11, 12]:

- Tension/compression: (for different tubes sizes) 13/10 MN, 25/20 MN, up to a max of 30 MN.
- Bending: 1 MN m/100 ft at 160 °C.
- Temperature up to: 360 °C.
- Internal and external pressure: 3000 bars
- Accuracy of tension/compression: ±1% of the measured value.

There are two types of loading equipment on the market:

In the first case (1) the pipe is stressed into a stiff rectangular frame by a load produce by a single large hydraulic cylinder mounted on the axis of the pipe, Fig. 8.2.

In the second case (2), two parallel cylinders mounted on both sides of the specimen, load the pipe, Fig. 8.3.

Loading for bending the specimen is produced, in both cases, by two cylinders acting in the transverse direction (Tables 8.1 and 8.2).

8.2.1.1 Functional Structure

Both previous testing machines have a common functional structure that can be developed at several levels. Figure 8.4, shows the first, second and third levels.

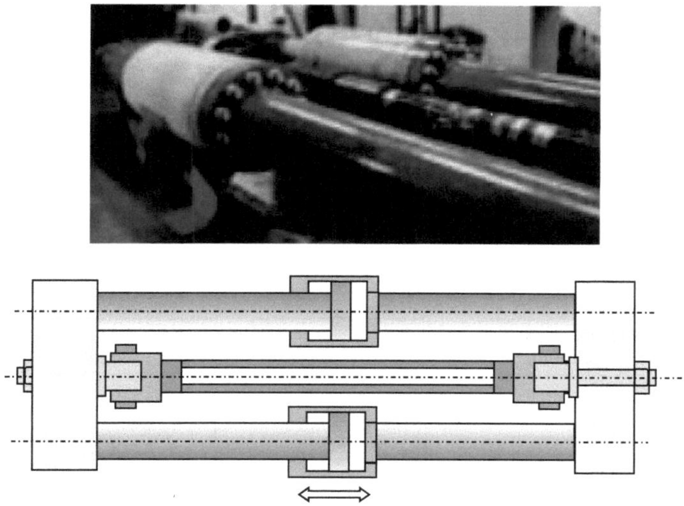

Fig. 8.3 VAM equipment and sketch [12]

Table 8.1 Analysis of the Pros. and Cons. of the solution in Fig. 8.2

Pros.	Cons.
• Simple very stiff frame	• Very large actuator with seal problems
• Only one actuator	• Difficulty in finding a large cylinder
	• High cost of the large cylinder
	• Long delivery time of the cylinder supplier
	• The chamber of the hydraulic cylinder for pulling the pipe is the smallest of the two chambers, due to the presence of the stem: correspondingly, the traction load is less than the compression one
	• Difficulty of transporting of the monolithic structure

Table 8.2 Analysis of the Pros. and Cons. of the solution in Fig. 8.3

Pros.	Cons.
• Simple frame	• Two simpler cylinders
• The chambers of the hydraulic cylinders for pulling the pipe are the largest due to the absence of stems	• Medium difficulty for finding two cylinders on the market
	• Medium cost of the large cylinders
• Corresponding traction load is larger than the compression load, as it should be	• Risk of different displacements of cylinders
	• Medium difficulty for finding two cylinders on the market
	• Cross-head must be very stiff to avoid bending on cylinders

(a) *Functional Structure of subfunctions of level 1*

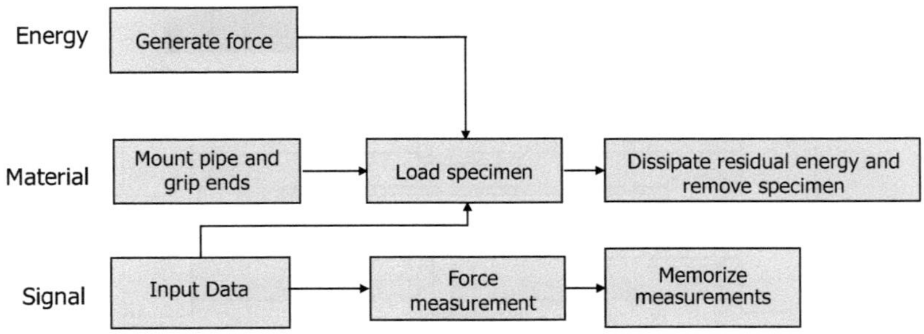

(b) *Functional Structure of subfunctions of level 2*

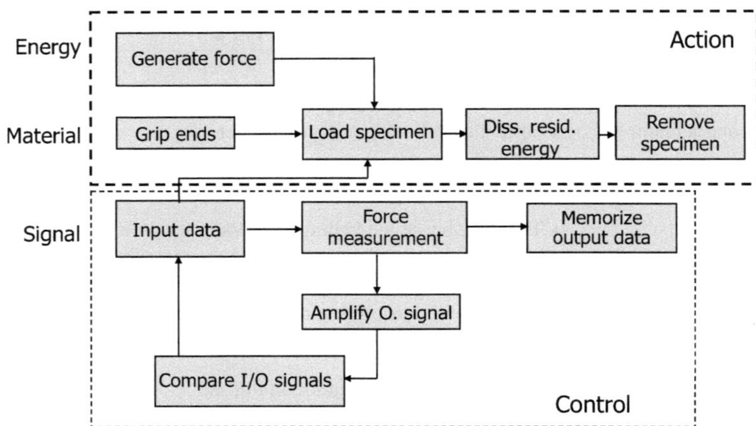

(c) *Functional Structure of subfunctions of level 3*

Fig. 8.4 Functional structures at different levels of details. Modified from [4]

It is important to note that the functional structure is independent of the technical solutions adopted in both cases. Consequently it is necessary to examine what type of solution have been adopted for every single sub-functions, if we want to express a judgment on the *pros and cons* of the two cases.

8.2.1.2 Steps of a New Design Proposal

You must note that the functional structure is built for solutions existing on the market and the method of comparing advantages and disadvantages represents a first step towards a real systematic design approach [10]. The method is less rigorous than the approach in the cited book, but this is the first step of training for critical thinking. An abstract search for general solution principles is a very complex task, suitable for advanced design experience.

This method is inspired by the cognitive psychology expressed in the book [8] that states its point of view in the title: *One Thing Leads to Another*, a clear reference to an evolutionary idea of invention, starting from what already exists.

The method is easier and usable in many cases and can be preparatory to the general, systematic method.

In any case, the most inventive phase of each method that decomposes the overall function in a variety of sub or sub-sub-functions is another one: to concentrate different solution principles on a limited number of elements.

The present case study highlights these points:

- Both pieces of testing equipment are present on the market, and substantially equivalent in technical terms.
- Before designing a new piece of equipment, in this case a new *conceptual variant* of the existing ones, we should carefully consider the *pros and cons* of existing design alternatives to satisfy the sub-functions described in the functional scheme.
- If a reasonable criticism highlights the possibility to utilize new alternatives, only after this analysis should they be applied to a new project. It must be accepted that new solutions can arise from new scientific developments, technologies, production methods, i.e. new techno-scientific knowledge and also from inventive intuitions. Not only: new solution principles can derive from new assembled groups offered *off-the-shelf* by new (or reduced) market availability, pushed by new political, commercial, logistic and economic factors, to say nothing of other constraints. The difficulty lies in the fact that criteria for judging a design of a new conceptual variant are numerous and difficult to predict in advance. A designer has significant expertise to evaluate the design options *ex post*, i.e. after the solution has been proposed and applied, more than the attitude to invent something totally new.[2]

8.2.1.3 Morphological Matrix

A visual tool to compare design alternatives used in the assembly of solution principles is provided by the morphological matrix, that combines principles for every critical sub-function, to help the designer come up with new solutions [9, 10]. It is

[2]*It is always easier to criticize (something existing) than to create (something new).*

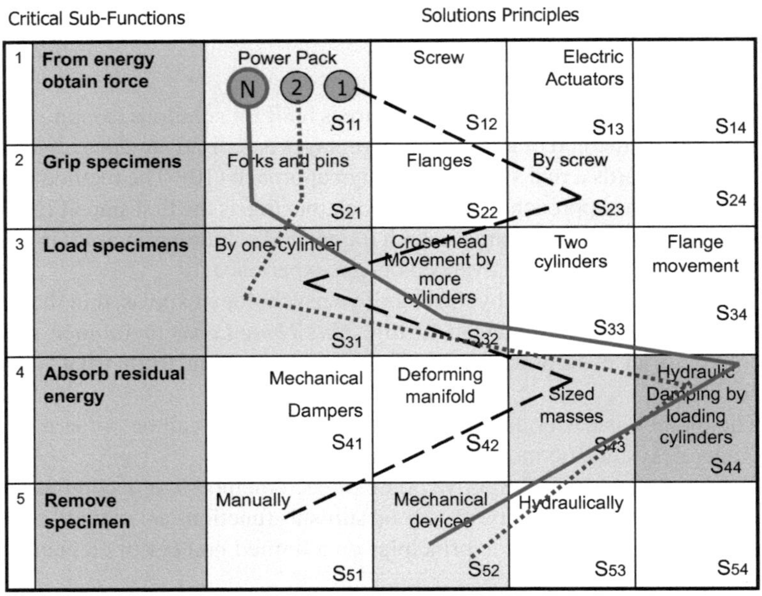

Fig. 8.5 Morphological matrix of three types of loading equipment. Modified from [4]

essentially a selection tool, to compare combinations of single solutions principles in several conceptual variants.

This technique was developed by Fritz Zwicky, a Swiss astrophysicist, based at the California Institute of Technology [13].[3]

The morphological matrix makes it possible to visualize then select several solutions principle combinations (that we can call conceptual variances), including new design options in ways you might not have thought of before. For the present case, a possible morphological matrix has the appearance of Fig. 8.5.

The combinations of the solution principles that give rise to three conceptual variants are shown in this matrix. # 1 is the conceptual variant of Fig. 8.2, # 2 is the variant of Fig. 8.3 and # is a new one that will be illustrated.

8.2.1.4 A New Design

The morphological matrix describes only the main technical solution principles that could be adopted for fulfilling the functions. The market limits the choices with

[3]Zwicky applied Morphological Analysis (MA) to astronomical studies and to development of jet and rocket propulsion systems.

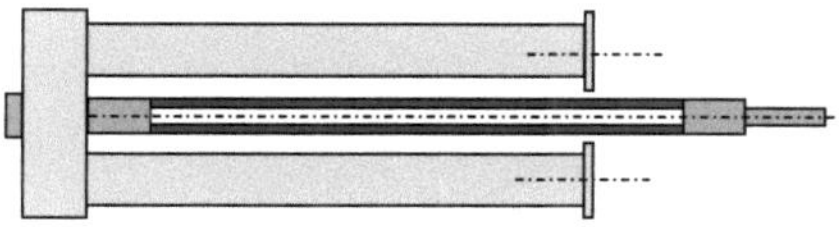

(a) *Pipe Specimen mounted on a C-Frame.*

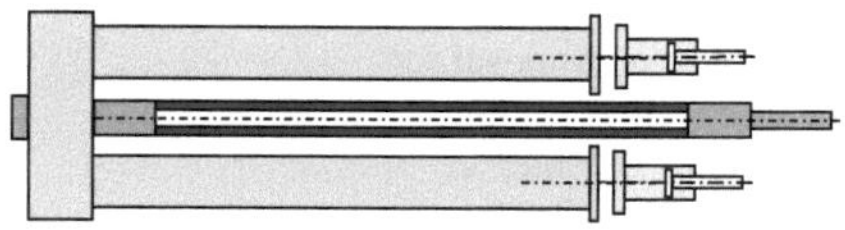

(b) *Addition of 4 Cylinders*

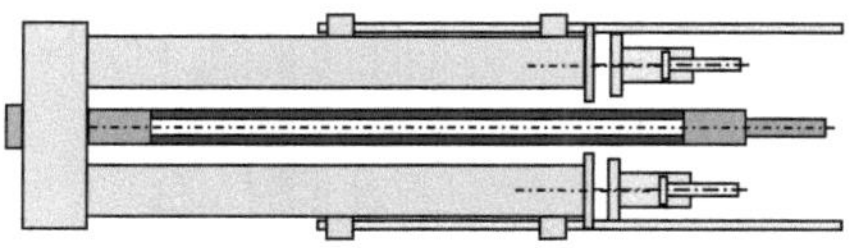

(c) *Addition of 2 guides.*

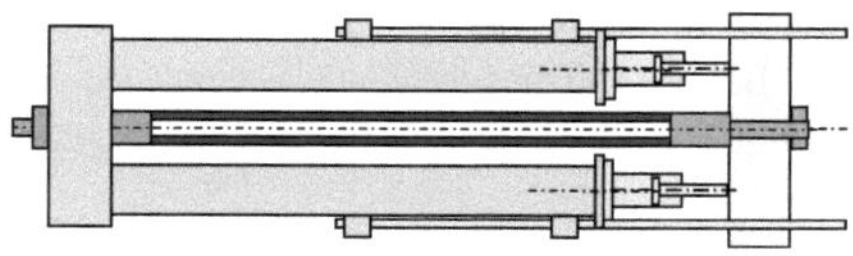

(d) *Addition of a cross-head*

Fig. 8.6 Assembly scheme of the variant *New*

other technical, logistic, and commercial constraints, which, unlike the initial designer's intentions, might suggest different solution principles, alternatives to the most obvious.[4]

In the present case, these of such constraints could be:

- Wide or limited availability of hydraulic cylinders of different capacities on the market.
- Wide or limited availability of other components such as chrome and ground guides, special high resistance alloy steels, etc.
- Opportunity to use a larger cylinder chamber for the traction and a smaller chamber (due to the stems of the piston), for the compressive action.
- Possibility of disassembly of the equipment in smaller parts to allow or simplify transportation.
- Principle of the division of Tasks for the components, with the advantage of function clarity (Different components for loading, cross-head guiding, etc.).
- Flexibility, e.g. possibility to update the load capacity at low cost (in fact, due to flexibility of a modular construction, that was really done in the present case, during the years of intensive use of this equipment, with an increase of 30% of its maximum loading capacity).

A new variant that respected these new market constraints and embodied some designer inventions was considered and shown in Fig. 8.6, derived from the corresponding principles combination in the morphological matrix of Fig. 8.5, with tag N (New).

[4]The present equipment was developed for the Materials Development Center, Castel Romano (Rome) where it was implemented by a measurement system by custom-made strain gauge load cells (Manufacturer: Giuliani-Forlí Italy).

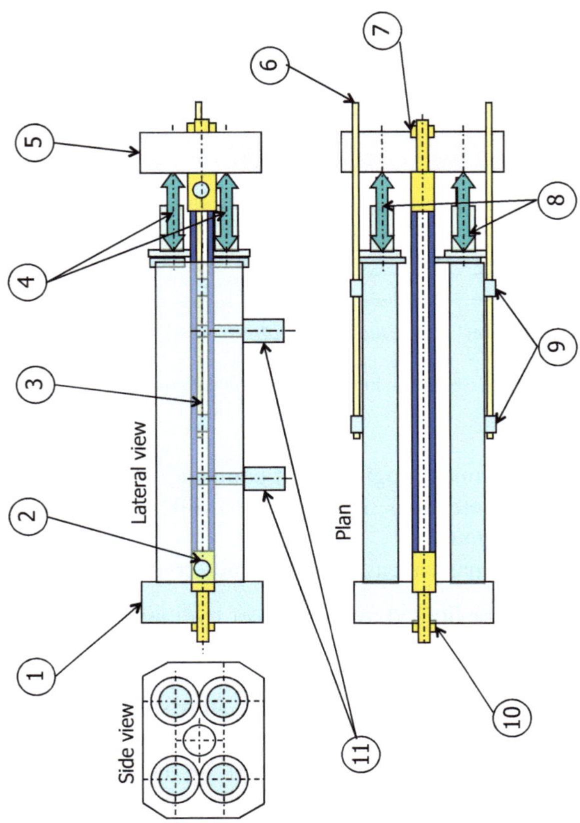

Fig. 8.7 New proposal for pipe-connection testing equipment 1: Frame, 2: Forks and Pins, 3: Pipe specimen with connection, 4/8: Four hydraulic cylinders, 5: Cross-head, 6/9: Guides and Bushing, 7/10: Bolts, 11: Transverse Bending

In order to clarify the concept of the machine, Fig. 8.7 shows the assembly operations of different components on the main C-frame.

8.2.1.5 Comparative Analysis

We remarked that the morphological matrix may not be sufficient to select the final design. Other kinds of selection criteria are needed to isolate the best conceptual variants and finally to indicate the most suitable one.

These assessments are made by a sort of *value analysis method* that uses tools derived by the logic theory for establishing the best solution. Through these tools it is possible to evaluate the solutions, associating a numerical value to each conceptual variant derived by the previous analyses. The value takes into account two factors, (as in the house of quality): an absolute evaluation and an importance coefficient.

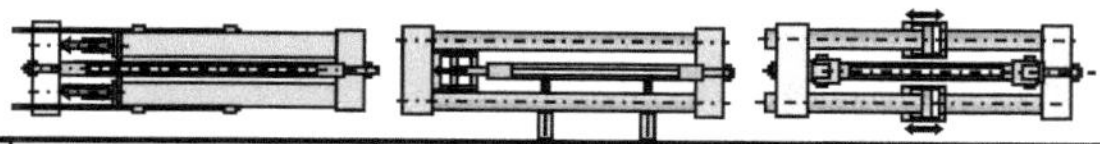

Technical Criteria	Importance %	Variant N		Variant 1		Variant 2	
		Value	Weighted Value	Value	Weighted Value	Value	Weighted Value
1. - Safety	14.3	10	1.43	10	1.43	10	1.43
2.- Ergonomics	8.6	10	0.86	10	0.86	9	0.77
3.- Producibility	14.3	6	0.86	6	0.86	6	0.86
4.- Controllability	8.6	4	0.34	6	0.52	6	0.52
5.- Assemblability	11.4	6	0.68	8	0.91	6	0.68
6.- Transportability	5.7	6	0.34	4	0.23	4	0.23
7. - Usability	14.3	10	1.43	6	0.86	10	1.43
8.- Maintainability	11.4	6	0.68	6	0.68	6	0.68
9.- Sustainability	11.4	2	0.23	2	0.23	2	0.23
Total	**100**		**6.8**		**6.6**		**6.8**

(a) *Relationship matrix for technical evaluation of three variants of testing equipment.*

Economic criteria	Imp %	Variant N		Variant 1		Variant 2	
		Abs. Value	Weighted	Abs. Value	Weighted	Abs. Value	Weighted
1.-Minimum Production cost	33	10*	3.3	5	1.7	7	2.3
2.- Minimum Maintenance cost	33	4	1.3	4	1.3	4	1.3
3.- Minimum Weight	33	2	0.7	1	0.3	3	1.0
Total	**100**	**5.3**		**3.3**		**4.6**	

The absolute value between 0 and 10

*many components on the market

(b) *Relationship matrix for economic evaluation of three variants of testing equipments. Maximum value for minimum cost.*

	Variant N	Variant 1	Variant 2
Technical rating	0.68	0.66	0.68
Economic rating	0.53	0.33	0.46
Overall rating (square mean)	0.60	0.47	0.56

(c) *Square mean between technical and economic rate, for the evaluation of three variants.*

Fig. 8.8 Value analysis of three variants for design of testing equipment. Modified from [4]

The mathematical product of the two rates represents a *a relative rating* for each criterion and the total sum of them an *overall rating* of the variant. Figure 8.8a shows an example of this first evaluation for the three pieces of testing equipment.

A similar analysis can be done by economic criteria,[5] comparing costs of production, maintenance, and weight, (considered as the most important factor of cost), for the same variants, Fig. 8.8b.

The square mean $R_{Total} = \sqrt{R_t \cdot R_e}$ [10] gives a mean rating of each concept variant. It compensates the two values better than an arithmetic mean. Finally, an overall rate R for each conceptual variant is given by the previous rating divided by 10, Fig. 8.8c.

The aforementioned systematic method, proposed in [10], describes this comparison with more details, and it should be consulted for further clarification.

The reader will note that, unlike the systematic method, the present method involves comparative analyses only afterwards, i.e. comparing the on-the-market existing solutions and the proposed new solution.

We believe that, in this case study, it is appropriate to carry out the analysis in the described way, summarized in the following steps:

1. Functional analysis of on-the- market existing solutions by means of more abstract descriptions of every function, sub-function etc.
2. Critical analysis of existing solutions, pointing out pros and cons. A new proposal would overcome the limitations of the original solutions. A usability analysis, as described in the example of the juicer, (in a previous chapter), can be a suitable tool. A new proposal must comply with specific constraints of various kinds, on technology, commerce, availability, maintenance, etc.
3. Comparative analysis between the new proposal and the existing ones on the basis of general constraints and of technical and economic criteria.

If a new combination of solution principles leads to strange variants, it means that review phases are necessary or a total reinvention and transformation of the existing solutions. This could be the way for development of innovative products (Fig. 8.9).

---

[5] Value analysis is a plan-based procedure for minimizing costs under the influence of comprehensive aspects. Calculated costs of the individual components are used to establish those costs that arise in fulfilling the overall function and the requisite sub-functions. This investigation produces important indications in the search for new solutions with worthwhile cost reductions [1].

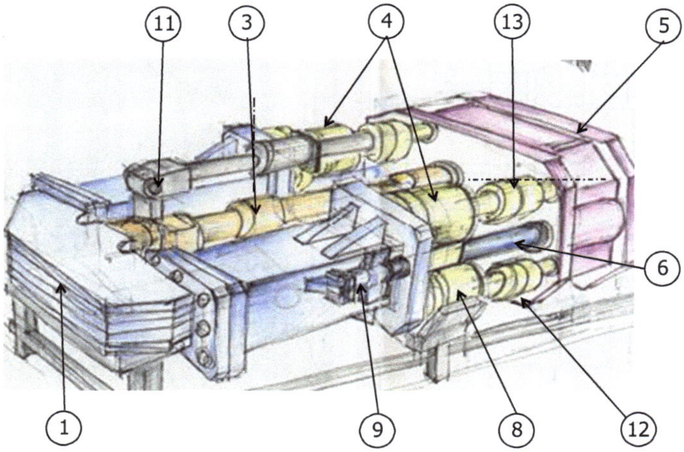

(a) *First hand sketch (A. Freddi) with the main modules. Numbers correspond to that of Fig. 8.7, with addition of # 12 and #13 hinges*

(b) *Prototype during the tests phase*

Fig. 8.9 The final prototype, designed by S. Curioni and A. Freddi [2]

8.2.1.6 A Comment on the New Solution: How to Avoid Statically Indeterminate Structures

The insertion of four spherical joints between cylinders and stems was made to ensure the transmission to the cross-head of only axial forces, Fig. 8.10. The main point in favor of the proposed solution lies in the detachment between the specimen and the machine from the point of view of hyperstaticity. In fact,

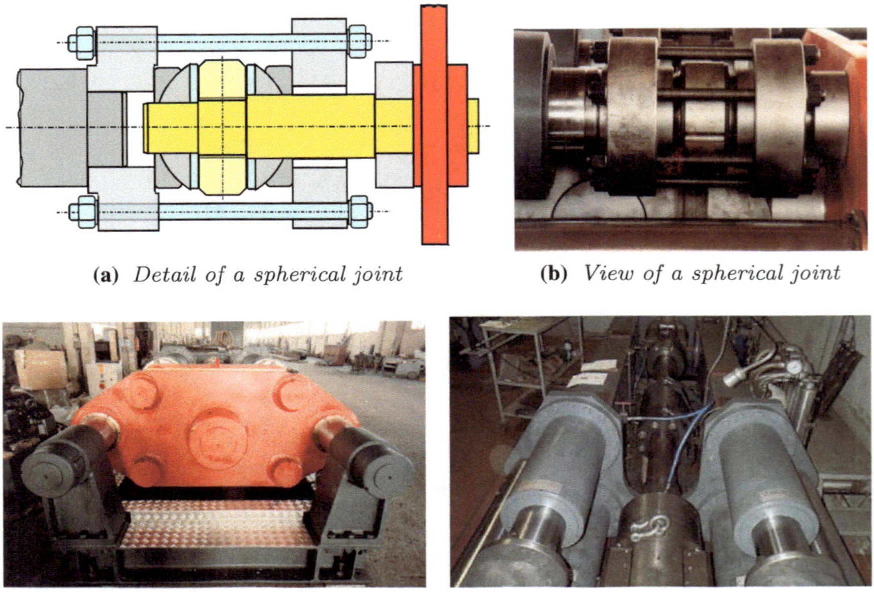

(a) *Detail of a spherical joint* (b) *View of a spherical joint*

(c) *View of the cross-head* (d) *View of hydraulic cylinders*

Fig. 8.10 Details of equipment for pipe-connections testing. Modified from [4]

small imperfections of alignment of the tube due to the machining tolerances, differences in the forces of the four cylinders in parallel and other stochastic variables difficult to quantify, maintain the load on the working axis that is fixed by the guides.

8.2.1.7 A Variant of the Previous One

Figure 8.11 shows an improvement of the Variant # N. This solution avoids loading the guides with the reaction forces of the side cylinders and closes the bending moment on the test specimen.

It is an application of the principle of the forces path reduction, see [1, 10] for a complete discussion.

8.2.2 Design of Fracture Mechanics Testing Equipment

On the same principle of avoiding statically indeterminate *hyperstatic* structures,[6] the design of equipment for fracture mechanics tests is based on a statically determinate structure.

[6]A structure is hyperstatic (or statically indeterminate) when the static equilibrium equations are insufficient to determine the internal forces and reactions on that structure, since unknown forces (or variables) are more than the equilibrium equations.

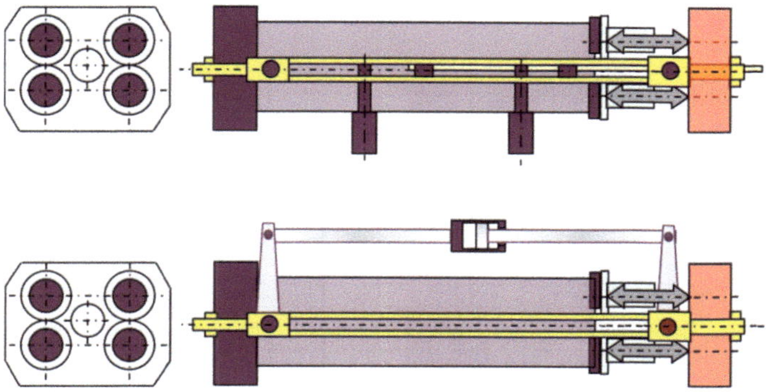

Fig. 8.11 A different way to impart lateral bending to the specimen (low side), compared to the previous solution (top side)

Specimens	CT
Dimensions	Thichness B up to 150 mm
Cost	Less than 50.000 Euro
Max capacity	3 MN
Fatigue load	20 KN

Fig. 8.12 Specification data for a testing equipment for fatigue pre-cracking and fracture toughness measurement in CT specimens. Other features are: elimination of external guides, insertion of energy absorption devices, and the possibility of static and dynamic use at low frequency, only for traction loads

Generally, the structure is planned very rigid and heavy to ensure the centering of the force on the specimen fracture plane. Conversely, in this prototype, the same specimen orients the load of the two hydraulic cylinders in a vertical plane by virtue of two spherical hinges, mounted on the front of the cylinder pistons. The specification list is shown in Fig. 8.12, while Figs. 8.13 and 8.14 shows the testing equipment.

8.2.2.1 Some Remarks on Production Technology

The main components are built from the following materials:

- The frame was designed joining together layers of four steel sheets, up to a total thickness of 300 mm.
- All the pins in low alloy steel.
- Cylinders in hardened steel.

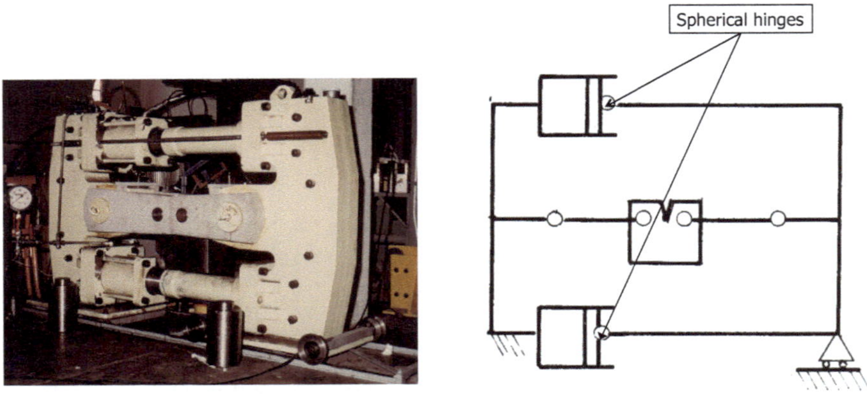

(a) *Picture of the testing equipment* (b) *Static scheme*

Fig. 8.13 Picture and iso-static solution of a fracture mechanics testing machine. Designers: A. Freddi and F. Persiani. Modified from [4]

- Balls in hardened steel.
- Dampers, made with tubes concentric to the tie rods, are made in annealed copper.

One of the positive side-effects of determinate structures is the application of the principle of the *division of tasks*, by which each function is allocated to a specific carrier. It produces unambiguity in calculations (clarity). Division of tasks is a principle of clarity [1].

Nevertheless, statically determinate (isostatic) structures, (when equilibrium equations are sufficient to determine all the forces and reactions on the structure), are not always the best solution.

8.2.3 Constraint Reduction for Testing Uniaxial Specimens

The same principle of constraint reduction is adopted in a device design for obtaining a clean uniaxial load for a specimen in a universal testing machine. Figure 8.15 shows a solution suggested among others, reported in [5].

The two rigid columns of the universal machine guarantees a push-pull loading of the lower mobile cylinder rod perpendicular to the lower head, but, if the specimen is constrained to the upper cross head, it is not possible to avoid a spurious bending moment along the specimen, due to a possible (small) relative inclination (amplified in the Figure) between the two cross heads and a possible alignment error between the grips of the specimen. The suggested solution is the following: a roller bearing with spherical seat is mounted that is free to rotate and translate along the surface upon

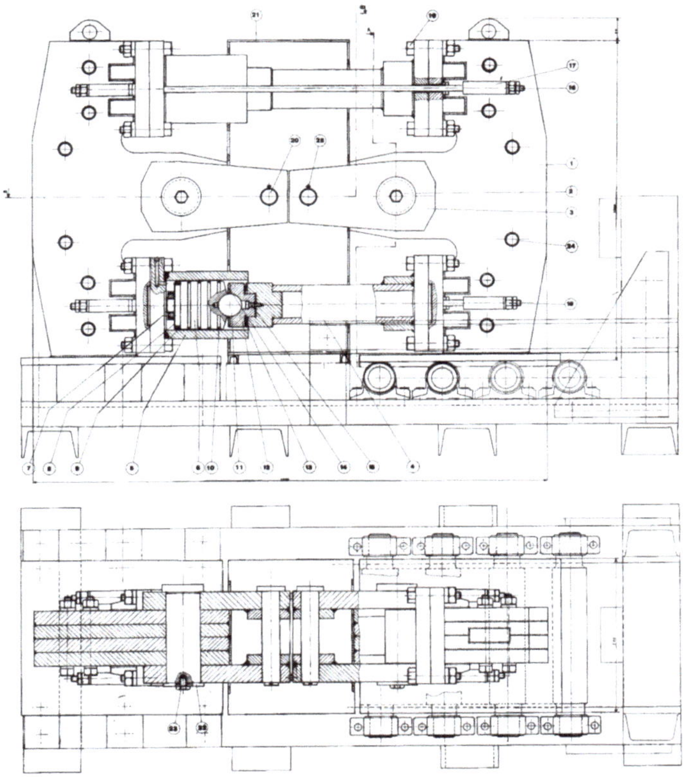

(a) *Drawing of CT Testing Equipment*

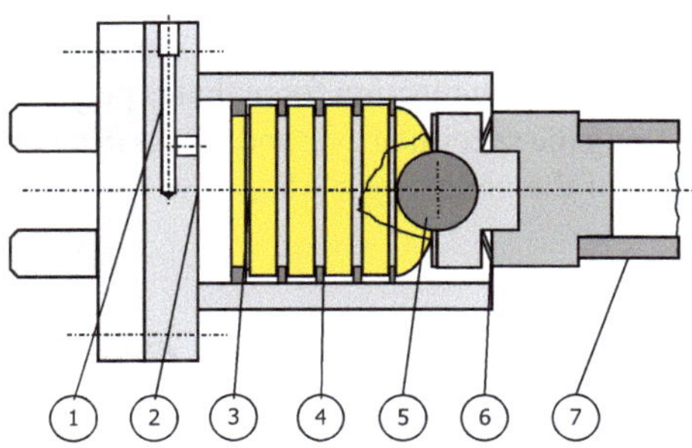

(b) *Hydraulic cylinder with a spherical hinge. 1:input pressure, 2: cylinder, 3:piston, 4: seals, 5. spherical hinge, 6:disc spring, 7: strut*

Fig. 8.14 Design of equipment for fracture mechanics CT specimens testing up to a thicknesses of 150 mm

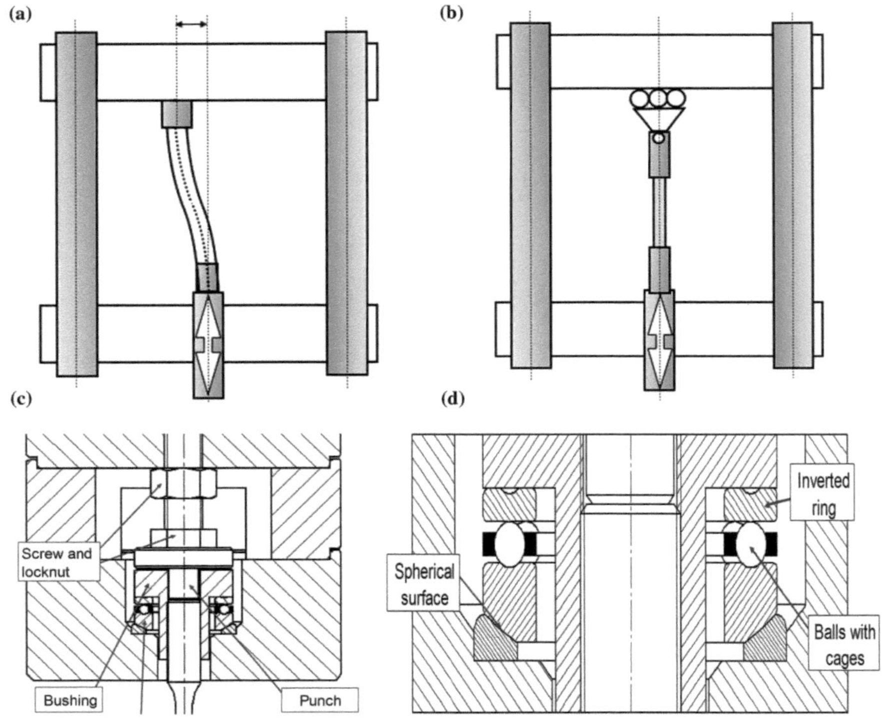

Fig. 8.15 A device for loading a uniaxial specimen avoiding spurious bending effects

which the roller rests. This is obtained *mounting the bearing with an inverted ring to realize a sliding constraint.*[7] In this way the bearing can absorb both misalignment and inclination errors. The resulting reaction force that is perpendicular to the mobile cylinder, is maintained along the specimen axis and works *in tension and compression* due to a special assembly of the parts, Fig. 8.15c, d.

8.2.4 Counterexample: Development of a Gripping Head for Torsion Testing Equipment

An example of statically indeterminate structures was introduced in the design of a grip in torsional equipment, for transmitting torsion moment to a specimen. This type of structure can be suitable for a principle of safety due to the redundancy offered by a hyperstatic structure.

[7]One of the authors is indebted to the late maestro E. Funaioli for this observation.

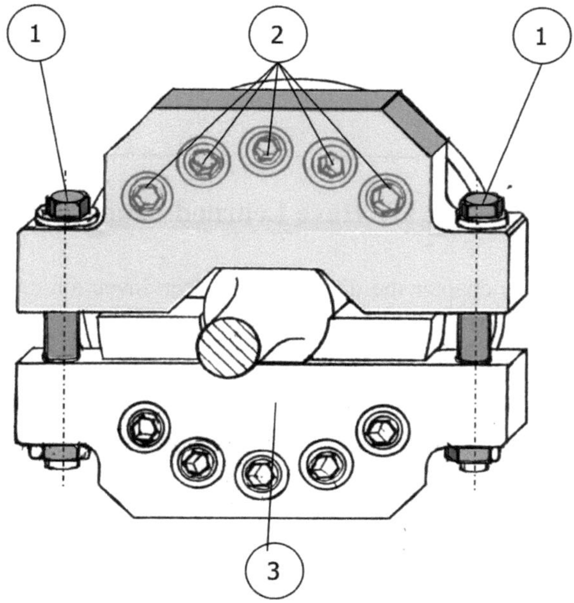

Fig. 8.16 Gripping head for laboratory equipment for applying alternating torque up to 25 kN m to specimens of 50 mm in diameter

To grip a specimen for variable torsion moment application is not a trivial task, since you have to combine the sub-function of a strong gripping of the ends of the specimen with the sub-function of applying variable torsion moments.

The solution that was introduced for the moving grip is shown in Fig. 8.16. The principle of statically indeterminate structures is applied. In fact, the clamping system is double and the block of the second tightening influences the forcing stresses created by the first tightening. Both must avoid even minimum relative movements between the specimen and the gripping head. The system is hyperstatic but effective. To understand the concept we can follow the assembly sequence:

1. Fasten the first part of the vice to the mobile head of the machine by the lower screws.
2. Open the mobile part of the vice and insert the specimen.
3. Put the mobile part of the vice (gray in the same Figure) on the specimen head and tighten the screws # 1.
4. Fasten the mobile part of the clamp to the mobile part of the machine, tightening the screws # 2.

Two different clamping systems are recognizable in this design: two vertical screws for a first tightening of the mobile part of the vice to the fixed part and a second series of bolts, in perpendicular direction, to join also the mobile part to the alternate rotating head. The movable part of the clamp must have holes with a suitable clearance in order to allow the transverse screws to pass through after fixing the first vertical ones.

This second operation modifies the tightening-stress distribution generated by the first screws in a way that is very difficult to preview and calculate, violating a principle of division of tasks, but the global effect is beneficial to avoid micro-movements of fastened parts.

8.3 What We Have Learned from This Chapter

In this chapter the difference between invention and innovation has been demonstrated by examples.

The design of testing equipment benefits from some inventive ideas, but which are only developed to the level of prototype building. Even for these one-off products, however, one must develop a full design, introducing all the concepts of functional analysis, morphological matrix and value analysis, in order to select the best variant that can be developed up to the embodiment design of one ore more prototypes.

Functional analysis and morphological matrices are utilized in a different way than the one recommended for the systematic approach: for products already built and on the market, to identify the strengths and weaknesses of existing solutions. The authors believe that the best use of these tools is more for a critical analysis of existing artifacts than for an evaluation of new ideas.

Critical analysis of existing products and morphological matrices are the starting point: new variants can replace old existing solutions, only if new ones are able to reduce weak points and provide some advantages.

The methodology is shown for cases of prototype equipment not for design of mass production. Nevertheless, it is possible to discuss methodologies and technical solutions in these examples.

References

1. Beitz W, Kuttner KH (1994) Handbook of mechanical engineering, vol 1, ebook edn. Springer, London
2. Curioni S, Freddi A, Caligiana G (1990) Experimental techniques in torsional fatigue testing. In: SEM (ed) SEM Spring conference on experimental mechanics experimental mechanics, society for experimental mechanics, Inc. 7 School Street Bethel, CT 06801-1405 USA, Albuquerque, N. M. (USA), vol 1. http://www.sem.org/
3. Ferroni L (2015) Bruno Munari e il metodo progettuale. http://www.luciaferroni.com/2morrowson/lucia/capitolo2.pdf
4. Freddi A (2004) Imparare a progettare, vol 1, 1st edn. Pitagora Editrice Bologna
5. Freddi A, Olmi G, Cristofolini L (2015) Experimental stress analysis for materials and structures. In: Stress analysis models for developing design methodologies. Series in solid and structural mechanics, vol 1, 1st edn. Springer
6. Kuhn T (1962–1970) The structure of scientific revolutions, vol 1, 2nd edn. University of Chicago Press. ISBN: 0-226-45803-2

7. MTS Testing Systems (2016) Coupling connection testing. https://www.mts.com/en/products/application/energy/petroleum-geothermal/tubing-piping-coupling-connection-testing/index.htm
8. Munari B (2006) Da cosa nasce cosa (One thing Leads to Another). Laterza Roma-Bari
9. Myrup Andreasen M, Hansen C, Cash P (2015) Conceptual design. In: Interpretations, mindset and models, vol 1, 1st edn. Springer
10. Pahl G, Beitz W, Feldhusen J, Grote KH (2007) Engineering design: a systematic approach, 3rd edn. Springer, London
11. Pinette Emidecau Industries (2015) Engineering and manufacturing of large scale turn-key mechanical machinery. http://www.pinetteemidecau.eu/
12. V&M Report OCTG Division Special Issue (2005) Creating the OCTG products of tomorrow. Progress report 6, Vallourec & Mannesmann Tubes—130, rue de Silly, 92100 France, http://www.vamservices.com
13. Zwicky F (1969) Discovery, invention, research through the morphological approach, vol 1. The Macmillan Company Toronto

Chapter 9
Design of a Packaging Machine: General Description and Conceptualization

Vero che, come forma non s'accorda molte fiate all'intenzion de l'arte, perché a risponder la materia è sorda,
Now it is true we all too often find his finished work belies an artist's purpose, when his matter is deaf and won't respond,
(Dante Alighieri: La Divina Commedia, Paradiso I, 129. Translated by J. G. Nichols)

Abstract This Chapter is devoted to the description of the conceptual design of an industrial product, the SASIB ALFA Packer (Sasib S.p.a., Bologna, now part of the COESIA Group, Italy), an automatic machine for the packaging industry, developed in the eighties, under the direction of one of the authors. The Dante Alighieri motto "It is true that, as a form is frequently discordant with the intention of an art, because its matter in response is deaf", reminds us that in any human activity a gap always exists between theory and practical implementation. Design of artifacts shows a deep gap between conceptual phase and realization inside a business oriented organization. While theory gives guidelines for the design of a product, industrial practice is product oriented, i.e. its goal is to "develop a new product" taking into account any related activity "from cradle to grave"; while theory takes into account essentially "physical" constraints, a real mass production must consider human and social constraints such as time, expenditure, human resources, customers, and stakeholders, psychology, etc.

9.1 Premise

The presentation of the design is done in the form of a story and the reader requires some clarifications.

1. About the storytelling style.
   - The description of the machine and of strategic choices linked to the time when the project (at the beginning of Information Technology and electronic actuators) is written in the past tense. In fact, conceptual

A. Freddi and M. Salmon, *Design Principles and Methodologies*, Springer Tracts in Mechanical Engineering, https://doi.org/10.1007/978-3-319-95342-7_9

analysis represents an action carried out in a past time, when all the great design choices have to be made, many of which refer to an environment of technological knowledge that is now obsolete.
Furthermore, all the comments that clarify the choices are shown in gray boxes.
- Vice-versa, the embodiment realization of the prototype is described in the next Chapter, in the present tense. Currently, in fact, the machine is still built and sold and the updates have respected the fundamental setting of the original phase.

2. About the method. There was no clear application of a *deductive method*, from a total abstraction from all existing solutions to deduction of new ones. Instead, we followed an *evolutionary method* based on the know-how of the company and improved by updated knowledge. The only reference points were the general phases described by the VDI guidelines and ISO 9000 standards.
3. The distinction between conceptual and constructive design is labile. In fact, it is intersected with the division of the machine into two parts:
   - The front side, functional to the process.
   - The rear side consisting of mechanical drives.

   A part of the conceptual solutions adopted (e.g. type of gluing) was defined later through prototype tests.
4. The most innovative aspect was the transition from an old rigid electro-mechanical concept to a conception with distributed actuators controlled by a central unit. This allowed great freedom of design. Even the controls, widely disseminated in order to avoid blocking of the fast packaging process, benefited from this centralized choice.

9.2 Design in an Industrial Environment

While in theory, design can be seen as a series of "decisions", in reality a large part of design activity is carried out by the search for an acceptable solution to a given "problem" i.e. design involves a never ending series of "problems to be solved".

Design is a selection of "compromise solutions" that simultaneously takes into account the "constraints" typical of any design phase. It is never the search for the "best", which could never be achieved, but a sequential activity for the selection of "good" solutions [3]. In fact "green grass" or "white table" condition are rarely present, if ever, in industrial products.

Design is "hyperstatic activity".[1] The challenge of any design, and mainly of the most innovative ones, is to overtake and bypass constraints one by one, and to reach the target.

The list of limiting factors or constraints in an industrial environment is rich, but just to recall the main ones, they are related to:

- *People*: this would be a long list of people related limitations such as hard and soft skills, experience and knowledge, collaborative attitude, "social" group behaviour.
- *Time*: in any global highly competitive market, time is as important as performance; design is a typical "moving target" activity and new product success is strictly related to its *time to market*. A few months delay, missing a presentation date may change a product from a leadership position to an "also run" one, see Chapter on invention and innovation.
- *Company related limits*: the product must be advertised, produced, and maintained by company people that are controlled by a general manager and/or owners. Each stakeholder has his/her own limits and boundaries that, inevitably, reflect on the design phase and, very often, conflict with product development management.
- *Task priority management*: resource allocation management is a key activity in any phase of a new product development. The dilemma by top management of whether to improve an existing product or to bet on a new future one often gives rise to delay and discussion that slows down the real design activity.
  Moreover, the day by day decisions must take into account the other activities inside the company to be performed. The division of responsibilities is clarified through an analysis that can use the same tools that have been previously seen for customer needs analysis.
  - Who? (does the task)
  - When?
  - Why?
  - Where?
  - What?
  - How?
- *Customer voice*: "the customer is always right" but it's hard to translate this sentence into real requirements and distinguish "personal opinions" from real "long term valid" needs.
- *Policies*: these range from applicable laws and regulations, to people's opinions and related behaviour.

[1]In statics, a structure is hyperstatic (or statically indeterminate) when the static equilibrium equations are insufficient for determining the internal forces and reactions on that structure, since unknown forces (or variables) are more than the equilibrium equations. Vice versa, the system becomes isostatic (or statically determinate) when equilibrium equations are sufficient determine forces and reactions on the structure.

In the sense, the word *hyperstatic* means having excessive or redundant constraints with respect to those strictly necessary.

- *Budget*: the main design parameter is overall allowed expenses. Any expenditure must be taken into account, from the starting activity to the "magical" "first customer delivery," when cash flow ends to plunge.

9.3 General Description of Packaging Machinery

This case study deals with innovative industrial instrumental product[2]: *an high speed cigarette soft packaging machine*.

The term "packaging machinery" refers to a very wide group of automatic machines that perform the functions required to pack any product to be sold to consumers or the business market.

The boundaries of this group of machines is relatively fuzzy in respect of other automatic equipment such as assembly machines, food preparation, quality control and inspection systems and there is a large overlap with other automatic machines classes.

The main characteristics of packaging machinery are:

- Special purpose design with very limited, if any, capacity to operate on different products. On the contrary, *machine tools* are conceived, in general, to machine any material from wood to titanium, etc. In fact, any packaging machine have a single design to handle a very well defined product.
- Capability to operate with "natural" products (vegetables, tobacco, soft cheeses, soft sweets, fresh or frozen vegetables, sticky liquids, liquors, etc.).

Each product has very specific, seldom undefined, *process rules* and each packaging machine is conceived around such rules. Of course, specific treatment is a "problem" that, once "solved" becomes a strength for the producer: many important companies are created on the basis of the packaging solution of a specific product such as tea, sweets, soft cheese, ... and cigarettes. The Design development can rarely be organized in a clear sequence of operations, due to the specific product that is dealt with, Fig. 9.1.

The main phase is based on experience, almost never formalized. The experience is often formed specifically for a product and remains constant in the development of new ones. It is particularly critical in natural products and, in any case, in products with high variability "from piece to piece". Typical examples are:

- Bottling of liquor-type products that make a mess.
- Blood treatment in bags (risk of emboli formation).
- Design of slides to drop salads, frozen foods into containers.

[2]*Instrumental good*: something considered as a means to some other product; i.e., an instrumental product leads to something else that is product.

Consumer or intrinsic product on the contrary, is definable in relation to instrumental product. Consumer or intrinsic product is something worthwhile not because it leads to something else, but for its own sake alone. It is not a means to something else.

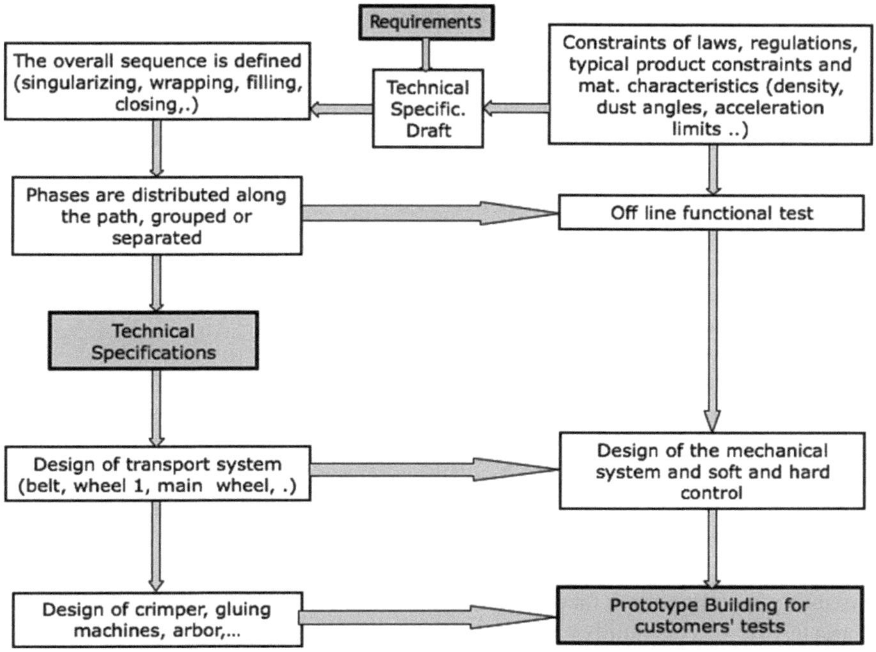

Fig. 9.1 Design sequence for packaging machines

- Packaging of biscuits: different from day to day, and the biscuits that must not be broken.
- Treatment of soft cheeses: only few companies know the secret of packaging very soft products.

9.3.1 Automatic Packaging Machine Classification

The "packaging process" is organized in a common series of functions that make up the overall function of the machine.

- Feeding the proper quantity of powder or bulk material.
- Separation and singularization of parts.
- Transport.
- Folding.
- Gluing and Drying.
- Assembly to form "transport units".

Automatic packaging machines may be classified according to different criteria.

9.3.1.1 The Purpose of the Overall Function

- *Packaged product type or status*: liquid, power, random bulk, solid elements (biscuit, chocolates, …), soft material such as paper rolls, tissues, …
- *Operation type*: primary protection layer, over-wrapping, product arrangement (layered, ordered or bag filling, …), boxing, …
- *Packaged product*: pharmaceutical, food, cosmetic, ice cream portion, …
- *Packaging material and shape*: paper, cardboard, film type, …
- *Combination with other operations*: product portioning and/or weighing, packaging material shaping such as cardboard box forming, blister shaping, bag production, "tetra pack" type forming, …
- *Cleanness level*: sterile, purity, safety, …
- *Automation level*: autonomy, speed, labour tending needs, …
- *Normative and applicable laws*: "Food and Drug", pharmaceutical codes, food laws, …

9.3.1.2 Flexibility

Flexibility is an important factor that makes it possible to change the packaged good characteristic such as number of pieces/packs, weight, size, wrapping materials, etc. Usually flexibility is divide into two class depending on the real user of the flexibility:

- Flexibility for the producer or modularity, i.e. the capability to customize the machine at the production stage by inserting optional modules or by changing product size. Typically, this flexibility is used to follow market requirements and to add optional functionality such as labeling, over-wrapping, …
- Flexibility for the end user, i.e. capability for the user to modify the characteristic of the packaged good, as for example, change number of cakes per pack, change the target weight, size, … cigarettes per pack, …

To enter the packaging world it may be useful to recognize another design feature: the product motion through the machine.

9.3.1.3 Type of Motion

We divide the machines into two classes depending on the type of motion performed by the packaged product: continuous movement or indexed movement.

1. *Continuous motion machine*: where the product, and its packaging material, has a continuous motion and all operations are performed without stopping the products. In this machine, the product is picked up by a continuous moving belt, or similar devices, and then moved to the end of the machine always running at an almost constant speed.
   Bottling units, cigarette making and filter application units, are typically made

following this principle. In general, this layout is used for machines with operating speed over 500 units /min (bottling machine run at over 1000 units/min).
2. *Intermittent, or indexed motion machines*. Most automatic machines are characterized by an indexed motion of the product through a series of "stations" where the packaging operations are carried out. The motion may be circular or linear or a combination of both. The tackle time of the machine is split into an *indexed movement time* and a *standing time* often in the ratio of 1:2. This type of machines is characterized by large indexed moving parts and one of the key components, as will be better described later, is the indexing mechanism such as the *malta cross* and *index type cams*.

9.3.1.4 Machine Layout

Machine layout may be of two types:

- *Balcony style*,
- *Mixed up type*.

Up to 1970, machines shared a common layout: the product flows on top or inside the machine or in front: machines are, typically, very compact.

After 1970, many new machines started to share a common layout: the so called "balcony" style characterized by a structure where the product flows along the front of the machine, while all mechanical driving systems are placed in a rear structure. The machine may be divided, conceptually, into two parts separated by a vertical plane:

- The front one easily accessible with the product handling system.
- A back part consisting of a sealed structure containing the driving system such as cams, indexers, gears, motors.

The two parts are ideally divided by a separating wall where shafts, plungers, … commands pass through.

On the contrary to the machine tool sector, for example, where modules are really *units* often available on the market (such as mandrels, guiding rails, safety gates, lubrication systems, …) in the packaging world there relatively few function modules that are ready *off-the-shelf*.

9.3.2 Particular Features of Packaging Machinery

- The majority of packaging machine factories build their product in small batches, say 10–20 typically, "to order," or also in very small batches.
- Very long life of products: products more than 20 years old are still sold and the machines we will describe were in production from 1984 to 2016.

- Production organization makes considerable use of sub-contractors for machining, surface treatments and casting.
- This supply chain principle gives rise to a very complex and intricate "local system" that has shown a very high level of flexibility and time responsiveness typical of the packaging valley.
- Market globalization is also changing the sector and, nowadays, all big companies have overseas subsidiaries for local consumption production in emerging market Countries like China, India, Brazil. The off-shoring from Italy is widely used for part machining in eastern Europe Countries.

9.3.3 Special Criticality Conditions of Packaging Machines

Special states of packaging machines are *Jamming, Stoppage and Restart* that must be considered with special design care since they are sources of important innovations.

9.3.3.1 Reliability and Jamming

Machine failures, i.e. the incapacity of the machine to operate because of breakage, part wear, … are relatively rare: MTBF has an order of magnitude of 10,000 h and is not a real problem for any stakeholder.

On the contrary, *jamming* is an event that happens many thousands of times: many machines have a typical time "between jams" of a few hours. As the word suggests, jamming has non typical behaviour or origin: any single machine has its own *jamming mode*.

Jamming mode may be originated by an almost infinite number of causes due to variability of: foil, paper or plastic, rolled up or broken ribbon, bad gluing, glue spread, breakage of materials like biscuits, product sticking, paper jam, etc. Jamming can also be caused by: paper moisture content, incoming product defects such as broken sweets, etc. Because of its random origin, jamming is often non repetitive but often happens in bunches and is typical of any single machine and any running environment.

The capacity to run without stoppage is often called *runnability* (as in paper mills) and is measured by the average time between two stoppages.

> Because of the "local" nature, i.e., influenced by the environment and local conditions (even atmospheric), actual values of runnability are never guaranteed, or even mentioned, by any machine producer.

9.3.3.2 Stoppage and Restart

Another typical aspect of packaging machinery that is worth mentioning the criticality of any stoppage and restart of the machine: any "emergency" stoppage due to any jamming may originate a subsequent difficult restart due to poor cleaning of faulty material, gluing problems, poor cooling or heating. The overall results is a considerable reduction of reliability which is manifested as great difficulties in restarting or, in the worst case, gives rise to a sequence of secondary faults, difficult to detect and dejam.

9.4 Packaging Machine Layout

Packaging machines are made from a very wide integration of elementary functional modules that perform the elementary operations needed to obtain the required performance of the machine.

Unlike to machine tools, these modules are not easily available on the market and, moreover, are not standardized in shape and function as they are designed specifically case by case and generally are never commercially available.

It is impossible to define any standard layout for packaging machinery and we may subdivide the components of any machine into three large groups:

1. Active "functional" modules that operate directly with the material to be packed or with the wrapping material, such as paper cutters, pick-and-place units, product pushers, etc.
2. Mechanical or electrical components that drive the functional modules such cams, index cams, etc.
3. Logic system to sense and control the machine operation.

For many years, components (1) and (2) were mixed up in the machine without a topological differentiation.

We can now consider the main functions referring to our case study.

9.4.1 Main Functions

The main functions of the machine are:

9.4.1.1 Transport Function of Product to Be Packed

- Belts or chains, accompanied by their containers.
- Pushers to slide the objects.
- Wheels with their proper containers.

- Chutes or rails to guide the product.
- Paper folders.
- Single foil extraction from a stack.
- Feeding from reels.
- Weight, integrity control, parts count.
- Filling of powder, solid or liquid.
- Robotic devices completed with grippers with different layouts (Cartesian, rotoidal, parallel, and "Selective Compliance Assembly Robot" type, etc.).

9.4.1.2 Handling Functions of the Wrapping Materials

- Reels shafts.
- Rolls to guide the ribbons of materials.
- Stacks of labels or blanks.
- Devices to extract a single label from a stack.
- Cutters.

9.4.1.3 Glue Application Function

- Smearing wheel.
- Spreading injectors.

9.4.1.4 Heaters

9.4.2 Mechanical and Electrical Components of Functional Modules

Each "functional modulus" is moved by a mechanical device such as:

- Continuous rotating shaft.
- Independently controlled motors of different type (AC asynchronous, DC, brushless, stepper, … linear, …).
- Pneumatic linear cylinders or pneumatic rotating motors.
- Brakes.
- Panels, bodies, windows, to limit human access, reduce acoustical emission, contamination, collect scraps, etc.

9.4.3 Logic System for Controlling

- Logic sensors (mechanical micro switches, proximity switches, optical switches).
- Logic software codes.

- Safety circuitries and/or software, doors locks, …
- Transmission and remote control units (Internet interface, M2M functions, …).
- Continuous output. transducers to measure weight, force, paper tension and thickness, …).
- Vision systems and related software
- Overall system control (PLC, computer, …).
- Motion generation and control.
- Pneumatic valves.
- HMI made by spy lamps, alarms, LCD panels, touch panels, buttons.

It is important to note that, while all machines contain some of the previous modules, they have a very different "embodiment" and cannot be separated: in a real packaging machinery several modules may be integrated to obtain two or more "combined functionalities" or they share a single driving device.

9.4.4 Cigarette Manufacturing

Cigarette production was, and still is, performed in large integrated factories, where cured tobacco-leaves are transformed into packaged cigarettes.

Cigarettes production factories have a very different layout but they share an almost "standard product flow", Fig. 9.2 describes the basic tobacco flow.

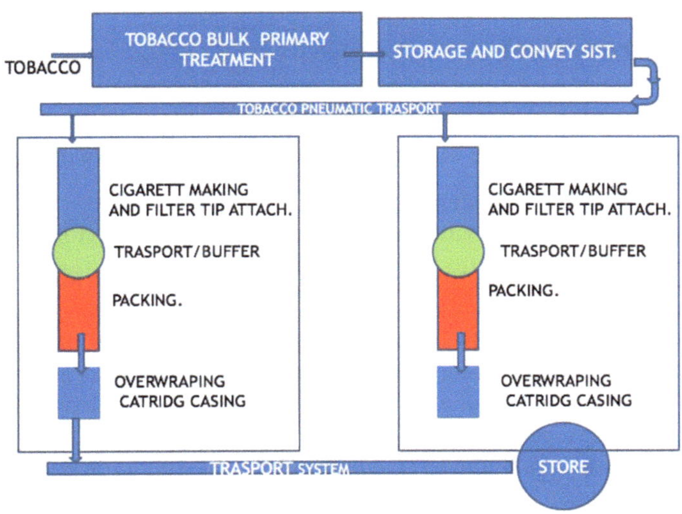

Fig. 9.2 Generic functional layout of a cigarette factory

At first, tobacco leaves undergo a so called “primary” treatment where tobacco is blended, toasted and cut into strands; these phases are carried out through an almost continuous flow that is often one single flow for the entire plant.

Downstream of this process, the tobacco is sent, by a pneumatic transport system, in parallel to a large number of “makers” i.e. machines that produce the cigarettes completed with their filter tip. The makers’ output is a multi-layer stream of cigarettes positioned on a special transport-storing system similar to a belt.

Cigarettes are then moved, with a continuous flow, to the packing machine placed alongside the maker often with a 1:1 direct link.

The packing group typically, consists of a *packer* that produces the pack linked downstream with a transparent film over-wrapping unit and, finally a machine that produces a carton, where a 10-pack carton is formed.

The 10-pack cartons are then moved, by belts, to casing and palletizing units.

When this story was first recorded, almost all modern factories were structured in this way and the product was very standardized in two types of packs as we will see later.

To give a quantitative idea of a plant in the 1970s, its main characteristics were:

- Running 24/24 hours for 7/7 days.
- Tobacco input 10–30 ton per day.
- Number of making-packaging lines up to 100.
- Speed of top class making-packing lines over 300 pack/min.
- Number of cigarettes/day: 10–40 million.

These data are given as an order of magnitude to better understand the plant characteristics.

9.4.5 Cigarette Packs

Cigarettes are sold “packaged” in a extremely large variety of packs but, for many years, up to about 1970, there were two dominant types of packs worldwide:

- “Soft packs” of paper and aluminum foil.
- Hard boxes made of cardboard and foil characterized by a “flip top” closure.

Both packs had 20 cigarettes: 20 was a real worldwide standard up to 1980 when the first 25 cigarette/pack was sold in the USA.

Soft packs were the worldwide standard with the exclusion of the UK were the hard box was widely used. Some countries had local type of packs as, for example, the shoulder box in Canada, single wrapping layer packs in Russia or France for the very local Gauloise cigarette.

The cigarettes too were almost standard worldwide with a few exceptions until around 1990: 7.80 mm in diameter and length of 80 mm for soft packs.

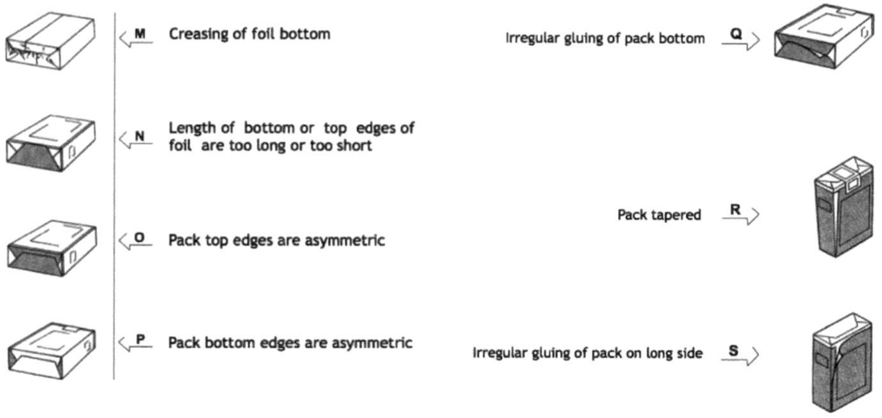

Fig. 9.3 Examples of unacceptable pack faults

The soft packs are made of three layers:

1. An inner layer made of aluminum foil 0.009 mm thickness.
2. A main wrap layer made of paper with the visible art work bearing the name and properties of the cigarette.
3. The paper pack is kept closed by a "stamp" that provides evidence of "not yet open" and, in some countries, is proof of payment of the government tax.
4. An outer layer made of transparent material sealed to protect the cigarettes from atmosphere.

Cigarettes: are arranged in three rows with 7-6-7 cigarettes each complete with their filter; their tobacco tip must appear dense and be sharply cut without tobacco protruding. The paper edge must be well cut without any "collar" mark; wrinkles along the body of the cigarette are absolutely unacceptable, See main defects in Fig. 9.3.

9.4.6 *Pack Quality Requirements*

The pack quality must follow strict regulations.

- It is important to mention that, in many countries, the cigarette is seen by the law as a food product and thus cigarettes and pack composition and products must respect many complex laws; in particular, in the USA the FDA (Food and Drug Administration) has always issued many laws on any material present in the pack such as glue, … or any material present in any form in packaging machinery. For example, fluorine based material, such as the commercial product Teflon, cannot be present in the machine even in hidden parts e.g. such as pipe connections), to avoid any risk for smoker of inhaling fluorine based smoke.

- The quality of the pack, like cigarette quality, was always a very important issue in a final market where customer loyalty is a key factor. Figure 9.3 shows examples of unacceptable pack faults; misalignment or "twist" must be less than about a few tens of a millimeter. We must first remember that quality was vaguely defined up to about 1980, when cigarettes producers and machine builders faced the problem as a primary objective.
- The size of the finished pack changes according to producer and depends on cigarette firmness and the shrinkage of cigarettes inside the pack: the average size is 81.5 × 53.5 × 22.5 mm.
- Moreover, there must be no tobacco particles visible in the pack.
- Packs must have their artwork properly positioned; in particular, the overlap must be visible for any skewness of the paper, it must have squared corners, flat sides, the folding must be symmetrical and well closed, the closing stamp must be placed symmetrically and in the middle of the pack.
- Cigarettes are very delicate products: in particular, high axial accelerations or any collision against solid obstacles will leave unacceptable marks on the cigarette tips that are the first thing that a smoker sees when he opens the pack.
- Packs must have a "life period" before been emptied and then there are other specific requirement such as:
- Packs must survive, keeping the cigarette as intact as possible, when placed in the shirt pocket of a soaked shirt as in the case of a smoker caught by the rain.
- Packs must survive without becoming loose when exposed to heat: some American companies called this condition the "Arizona dashboard test" meaning that the pack must not become loose when placed on the dashboard of a car in the Arizona desert: we will see later that the requirement raised difficult problems in ALFA design stages.

Other quality factors linked with the transparent film over-wrapping are not considered here.

9.5 Conceptual Design of the New ALFA Packer

The present example describes a product, the ALFA Packaging Machine produced by an industrial company. The reasons for this choice are the following:

- The ALFA is relatively complex in comparison with most packaging machines: it consists of 5,000 parts.
- The machine uses mechanical, electrical and software components.
- The story to be told is 40 years old, from 1979 to 1984, but the machine is not yet obsolete and is still produced and sold today.
- The machine was, at its time, a very innovative and, let's say, many solutions are still valid and, partially, unchanged today.

- Today all patents have expired, about 150 machines are in service, the machine will no longer be improved and so there are no secret or confidential information to be kept.
- In spite of its age, the machine is still a typical example of a high speed (500 pack/min) integrated mechatronic design.

Let's summarize the main ALFA features:

- Packaging 20 cigarettes in soft packs at 500 pack/min speed.
- Independent, computer controlled stepper motor for material handling.
- Multi-computer, 8 bits, control system.
- CRT Human interface.
- Start of design activity January 1979.
- Final prototype test: spring 1984.
- Design team of about ten persons; (mechanical, electronic, software, in support in different design phases).

The new owner of SASIB found a company situation that may be summarized:

- SASIB was producing the "IS 5000" a 250 soft packs/min packaging machine and developing its speeded up version, the "IS 6000", with a target speed of 300 packs/min.
- The main competitor, the GD group, was selling, very successfully, a 360 packs/min machine.
- SASIB market shares had dropped to less than 30% from an almost 100% of a few years before.
- SASIB had reduced R&D activity with a rather poorly motivated management.

9.6 Strategy and Product Planning

The new owner, involved directly in company management as an executive president, defined the main goal of the company:

To develop a soft pack cigarette packer capable of overtaking the GD competitor model. No budget limitation was defined and a three-four years time period was scheduled.

The new R&D management had to make a first key decision on product design strategy:

1. To revitalize a *continuous motion machine* already at the running prototype stage, designed to run at 400 packs/min but, de facto, never run.
2. To evolve the SASIB traditional machine, making a new version with a new target of 360 packs/min.
3. To start from scratch and develop a completely new machine using, possibly, the same conceptas the old SASIB machine.

Fig. 9.4 The SASIB IS 6000 packer running at 300 packs/min. Note the difficult accessibility and the compactness of the machine. Designer: A. Pratelli

It was not an easy decision because the SASIB management had lost any real decision capacity after having been "on sale" for many years, and the "crew", that was and would be the basis of any SASIB activity, would have preferred to follow old path and not to start a totally new project.

The new management started with the following decision:

1. To scrap the never-running continuous motion model.
2. To make a further improvement to the traditional SASIB packer, inserting some electronics but keeping the mechanical layout and defining a target speed at 300 packs/min, under the name IS 6000.
3. To develop a completely new machine under the name of ALFA to indicate the starting up of a new product line, Fig. 9.4.

In retrospective, they were wise decisions:

- The continuous machine was too advanced for that time, 20 years earlier than any industrial product of that type, and it was impossible to make it run at laboratory level too.
- The IS 6000 improved with new "Olivetti style" electronics became a successful intermediate product and 70 units were sold to the US and, later, hundreds to China.
- The ALFA packer, in spite of many difficulties, became a good product and, overall, 150 units were sold over for 40 years!

The IS 6000 main improvements were made by inserting electronic components such as variable speed, motor, optic sensors, electrically operated clutch, redesigned noise-deadening bodyworks.

Competitors Analysis and Innovation Contents

At the time of design-start, the market reference machine was the GD model $X1$ characterized by the direct wrapping over the cigarettes without any protective shield: this solution was different from the SASIB traditional solution that allowed a simpler mechanical design but with limited quality. The GD machine was also characterized by pack movement that was always perpendicular to the cigarettes, thus allowing a shorter travel in comparison with the SASIB tradition; last, GD made extensive use of the "toothed timing belt" a relatively new system. (First automotive use was in the early '60s, but commercially available from the early '70s).

The main advantage of the SASIB process was the quality of the pack. On the contrary, the process is slower because cigarettes must travel, with three movements, in a longitudinal direction.

Another innovative proposal was the glue to be used. This aspect was investigated in more depth in subsequent design phases, since because of its innovativeness, it was possible to delay its introduction to a more advanced phase of the project, in order to be able to test it directly on a prototype.

A few minor innovations were also introduced in the conceptual phase such as:

- Stepper motors.[3]
- A completely new electronic control system development based on a plurality of single board computers.
- CRT use as human interface.[4]
- Welded construction use as much as possible, replacing SASIB's traditional cast-iron solutions.

R&D Groups Organization

The R&D group at SASIB was traditionally organized in a strong hierarchical structure in five groups:

1. Maker, cigarette production, continuous motion machine design.
2. Packer and over-wrapper machine design.
3. Prototype test and measurement.
4. Electrical group.
5. Electronic and software group.

Before starting the project a new organization was set up, based on separation of "urgent short term continuing activities" from "important long term innovative" task:

- Continuing engineering mechanical group to follow all existing products.
- New product development with mechanical, electrical and software skills.
- Prototype construction, assembly, test and measurement.

Later on, a new group was created for CAD[5] and calculation activity.

---

[3] A stepper motor is a brushless DC electric motor that moves in discrete steps.

[4] Cathode Ray Tube technology, pre-LCD (Liquid Crystal Display).

[5] Computer Aided Design.

Two small teams were created inside the *new product development group*:

- A *mechanical design team* made of senior SASIB employees with a half a dozen very skilled draftsmen, capable of drafting intricate and complex units, with over 15 years of specific experience, under the direct control of a technician from "Olivetti".[6]
- An *electronic design team* with experience in microprocessor assembly software development, Intel 8080 at that time, and with a graduated engineer with experience in motor control, also from Olivetti.

The prototype test and measurement team was st up to build and test the product: the testing was led by a person with experience in mechanical measurement, noise analysis, and high speed filming (on chemical film at that time).

The mechanical workshop consisted of a few blue collars for assembly and testing activities and to machine urgent components. This group was also responsible for checking the parts machining that was subcontracted outside the company.

All the teams were "small in size" and "had joint responsibility" without any intermediate layer.[7] The SASIB top management followed the product development through monthly formal meetings. The design, and the other operations were not programmed, given the high uncertainty of almost every component.

To complete the scene it may useful to remember that at that time:

- There was no CAD available; (it was available only after the ALFA project release), and all drawings where hand made.
- FEM was used at University level only.
- The number of graduate engineers in the team was small. Most were undergraduates but with significant specific experience.

This was the "initial condition" of the design process, which lasted for about five years of "trials and errors", "progresses and failures", up to the prototype successful running at the customer's premises [2].

9.6.1 The First Model

Figures 9.5 and 9.6 show the first wooden model. The model was built in the woodworking department of SASIB[8] The wooden model, simplified in certain parts, is a little more compact than the final version. Another difference is the compression

[6]Elio Pagella the head of mechanical design. Olivetti S.p.a. was one of information technology players with a century-long tradition of research and innovation.

[7]The teams were managed by Mario Salmon (with the help of E. Pagella) without a formal structure.

[8]In those years the industries were strongly integrated and also possessed carpentry departments for packaging and models: the additive manufacturing was still a long way away!.

Fig. 9.5 Prototype and of the first series without body and with a glass shell realized at the beginning of the design development

Fig. 9.6 Details of internal parts of a wooden model

turret, here in the 8-stations version while the first version would be with 4 stations and would again become 8 stations, a decade later.

The last version of the 2017 ALFA packer is shown in Fig. 9.7: it differs somewhat from the original version produced in 1984. Over almost 40 years of life many modifications and improvements have been made but the general layout and design principles were kept. Continuous improvements caused a redesign of practically all the components of the machine but without a real change of their functions. The experience gained through the running of the 150 units also led to the introduction of many local small improvements: but, again, such improvements have not significantly changed the machine design. The main modifications introduced were:

- Complete redesign of the machine body: the original model had no full covers but only some protection for safety reasons.
- Redesign of the control system: the original one consisted of a series of Single Board Computers from Intel while the current model is controlled by a PLC from Siemens.[9]

[9] A Programmable Logic Controller (PLC) is a computer designed to work in an industrial environment that continuously monitors the status of devices connected as inputs.

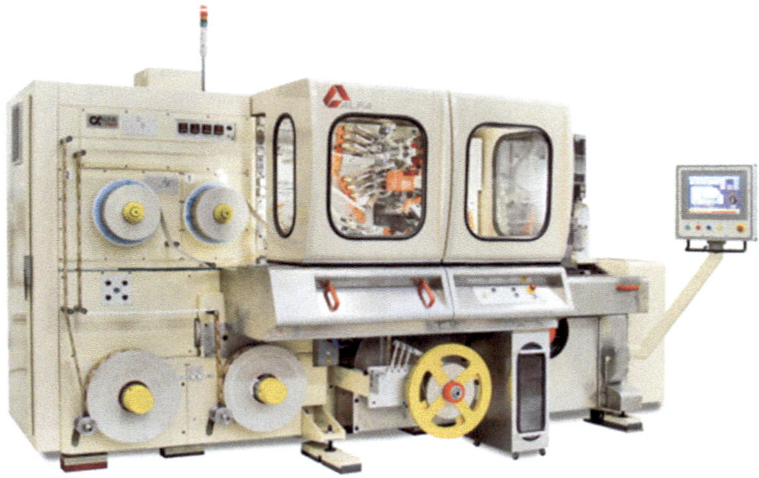

Fig. 9.7 The last version in 2017. Industrial design inspired by A. Pratelli

- Another very important modification was made to the motion timing. All movements timings were reviewed, obtaining large improvements for noise, smoothness, energy.[10]

It is appropriate, in describing the layout of the packer, to divide the machine in two parts, separated by a vertical transverse plane: on the front side are the flows of the product i.e. cigarettes, paper, … packs, and all the *active* components of the machine that performs the packaging operations i.e. feeding material, glue, cutting paper ribbon, etc.

On the back side of this virtual plane is the closed welded steel structure containing all the *commands* for the active components, Fig. 9.8.

Command components are the typical kinematics units used in packaging machinery such as: index cams to drive indexed motion shafts, plane cams to drive alternate movement components, gears, timing belts, etc. In brief, the two main parts are:

1. The *Packaging process* and related components.
2. *The mechanical part* moving the front parts.

This organization was adopted not only for ease of reading but also to enhance two different design activity phases:

(a) First, the design defines the parts that operate directly on the material to be handled such as paper, cigarettes, …
(b) The design must then develop the more traditional mechanical units for moving the "active" components.

[10]For sake of clarity, all figures, if not declared, were supplied by the COESIA group and shows the machine as it is produced now; the electrical control system described is the original one.

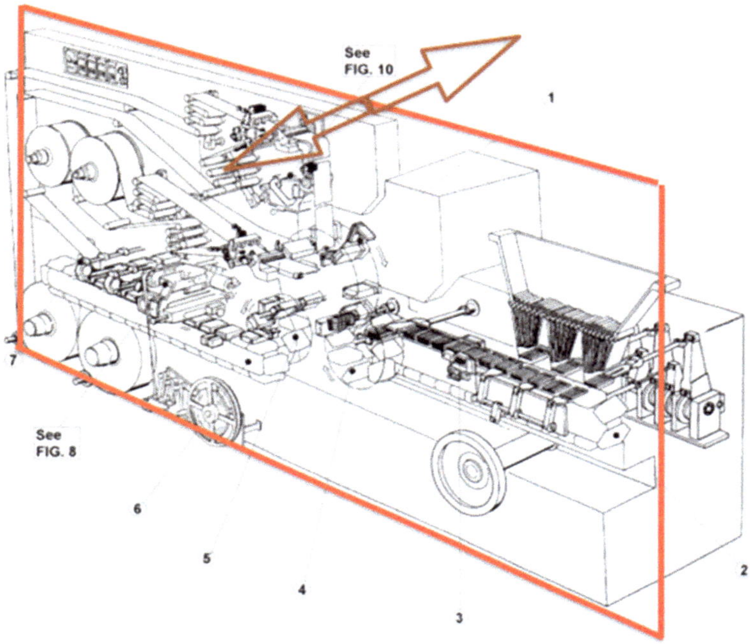

Fig. 9.8 Machine layout in balcony style: two parts ideally divided by a plane surface

Actually the difference between phases (a) and (b) is more logical than temporal as the two phases are broken down into sub-functions, each one consisting of a *process design* and then a *mechanical drive design*. According to the spirit of the book, the key part of a machine design is the conception and functions realization on the product: feeding, singularization, … gluing, … folding, etc. by means of appropriate solution principles. This is the most important part of the design where all the highly specific knowledge of the company must be put into practice and the key decisions are taken.

The two parts of the machine are very different from the point of view of the function and physical aspect:

- The performance of the machine, top speed, quality, runnability, human interface, … are defined almost completely by the *front part* of the machine that operates on the product, is easy accessible, and is not lubricated.
- The design of the mechanical part, let's call it *backstage*, is relatively easy because the required functionalities are clearly defined by the first phase output.
  The mechanical parts (enclosed in a steel structure, lubricated and not accessible) has a secondary role on the machine performance. In practice, the mechanical parts may break down, wear out, generate vibration and noise but, in our experience, all these "failures" may be spotted and eliminated by good design team. On the contrary, it is very hard to design, test, and correct, e.g., a "pusher" that must fold a label by 90° in about 0.5 ms!

This arrangement generates the so-called "balcony" type layout. This layout offers easy access to the product and a clear division between lubricated "dirty" parts and the "clean" front area, to be kept absolutely uncontaminated by lubricant.

9.6.2 General Layout and Reliability Requirements

A conceptual analysis of the mechanical driving units, typical of similar problems solving, in the tradition of the same company, highlighted a series of general requirements:

- Mechanical machine life: over 20,000 h at full speed running.
- MTBF of mechanical components: over 10,000 h.
- No use of independent computer controlled motors in positions with a collision risk.
- Use of independent motors only in positions where any mistiming cannot give rise to machine part collision.

In fact, the mechanical parts are characterized by a large number of "near miss" components and, in general, by many parts operating with risk of interfering movements; therefore the correct timing of any component is of paramount importance: a small phase error could lead to collision of parts that would cause the practical disruption of many parts.

A look at Fig. 9.9 where the timing of all cams and indexes is shown, gives an idea of the criticality of phasing.

The conceptual analysis of these units suggested dividing the machine moving parts into two groups:

- Mechanically connected parts with a single "degree of freedom".
- A series of part electronically connected, moved by a series of stepping motors.

Other informal principles derived also from the company know-how are:

- Not to use cast-iron technology for the machine main body.
- To use cast light alloy for any moving part, as a general rule (to reduce masses, thus inertial forces).
- To use forced lubrication for all parts.
- To buy index cam units.
- To use desmodromic cams and consequently, to reduce the use of springs.[11]

A final disclaimer: it is true that functionality is the real "design driver" and that, conceptually, the mechanical parts are designed after definition of the front part but,

[11]The reason is different from that presented in another chapter on the control of the vibrations of the springs in the distribution of the valves in an internal combustion engine: in this case, the choice of desmodromic cams allows the optimization of the laws for the forward stroke and for the back stroke. However, the choice makes use of a current experience in the area (especially for the technological problem of processing and mechanical treatments).

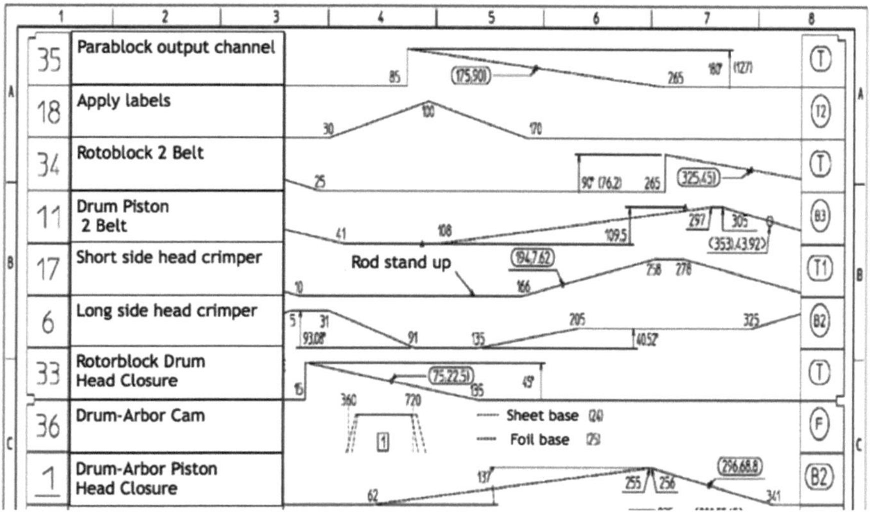

Fig. 9.9 A part of the timing chart of the machine: the chart shows 9 over a total of 34 lines corresponding to the degrees of freedom of the machine, both mechanical and software driven. Each degree of freedom, such as pusher indexer, etc., has its own specic phase carefully tuned to reduce acceleration and improve product handling

in reality, the design must be carried out in parallel because, as in all design activity, the final solution is always a compromise between opposing requirements.

The applications of new rules was not always easy because they forced big changes in the whole company, obliging, for example, the buying department to look for new sub-contractors and new "input good quality control" way of operating.

The cost of the machine was never taken into account at the design stage. After the prototype completion, SASIB had a new CEO who was concerned by the production cost of the machine that was almost a double the previous model: a special investigation was made showing only a small cost reduction possibility and the subsequent decision was not to activate any cost reduction for the entire project but only for a few sub-assemblies. Some cost information is given at subsequent subsection of next chapter.

9.6.3 *Front Active Side*

The ALFA packer design development will follow this subdivision: at first the active front, and later the driving part.

9.7 Requirements on Packaging Functions

The machine input materials are:

- Cigarettes in a mass flow chute.
- Aluminum foil in two reels.
- Wrapping pre-printed paper labels in two reels.
- Fiscal stamps in a reel.
- Hot melt type glue as solid chips.

The machine outputs materials must be:

- Completed packs.
- Rejected packs.
- Rejected cigarettes.

The produced packs must consist of 20 cigarettes, wrapped with two layers:

- An inner layer of aluminum foil.
- An outer layer made by paper with artwork printing.
- A third element, called a stamp, keeps the pack closed and grantees the closing of the pack.

The wrapping material[12] is fed in reels with the following main characteristics:

Aluminum Foil

- Diameter max 20 mm
- Width 110 mm about

Paper Reel

- Diameter max 600 mm
- Width of about 96 mm

Stamp Reel

- Diameter max 520 mm
- Width 20 mm.

Paper and stamps have optic marks with a special mark at the material end to trigger automatic splicing and to change feeding from the finished reel to the other new reel. The ALFA packer functions and the related material flows are shown in Fig. 9.10.

The glue has a non-secondary role and can be chosen by the manufacturer, with certain external constraints of a hygienic and health nature. The first research of the company was on the possibility of using non-conventional glue, compared to that used by the competition and by SASIB itself in other machines: a hot melt glue instead of the almost standard cold type casein based glue.

[12]The packaging product must meet the precise technical requirements (that can not be modified), set by the companies that sell the product.

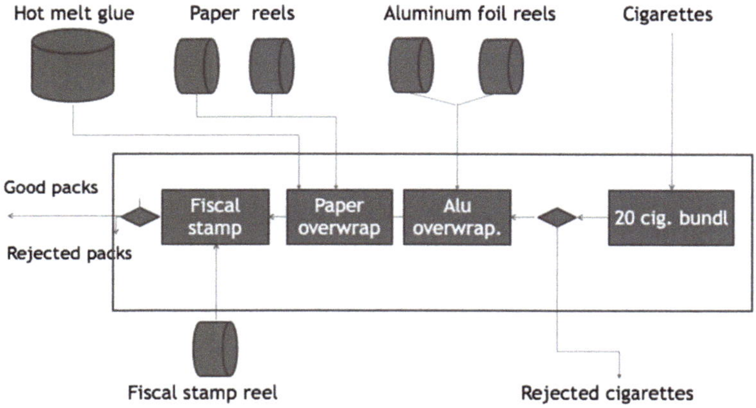

Fig. 9.10 ALFA packer functions and material flow

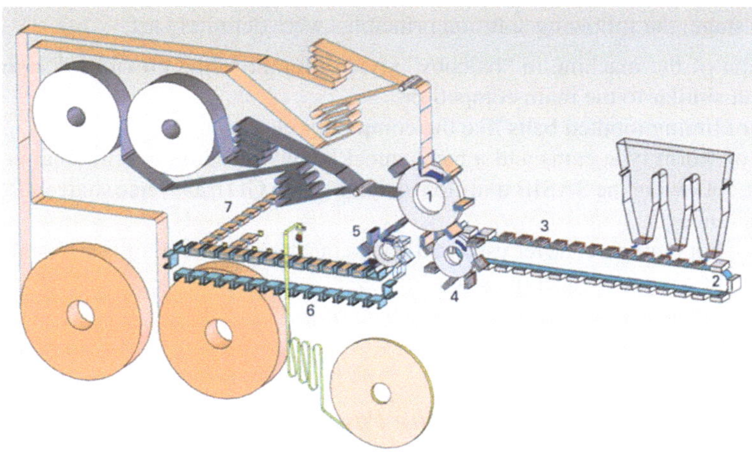

Fig. 9.11 Material flow: (2) Boxes on the first belt, (3) First belt, (4) Compressor drum, (1) Arbor drum, (5) Third drum, (6) Second belt

The packer is shown in Figs. 9.8 and 9.11, where:

Cigarette Feeding Hopper

1. *Arbor Drum* where the materials are wrapped around the cigarettes.
2. *First Belt* to place cigarettes in place 7-6-7.
3. Cigarettes count checking.
4. *Compression Turret* to compress the cigarettes.
5. *Foil folding drum.*
6. *Second belt.*
7. Good packs exit.

In Fig. 9.11 we also have:

- Two rolls of wrapping paper
- Two rolls of foil
- One roll of fiscal stamps.

9.8 Solution Principles

The next step was to select the solution principles in order to realize the functions, see the operational sequence in Fig. 9.12.

9.8.1 Solutions Adopted for the Main Functions

At this stage, the following solution principles were definitely set:

- Design of the machine in "balcony" style different to the previous design, with layout similar to the main competitor.
- Use of timing toothed belts like the competitor choice.
- Use of worm type cams and a mechanical layout based on longitudinal oriented shaft, following the SASIB tradition and unlike the GD transverse shaft and "Malta cross" indexer.
- Use of hot melt glue, never used in tobacco machines, instead of traditional casein cold glue to reduce machine size.
- Use of desmodromic cams to avoid any springs.

9.8.2 Solutions Adopted for the Packaging Function

Once the main key decisions had been taken the following step was to define the solutions for the "packaging process," i.e. the single activities starting from the mass of cigarettes (falling down from the chimney-like chute) to the finished pack.

To capitalize the SASIB experience, the process was designed very similarly to the traditional one. The functional sequence was designed along two convergent lines, one for cigarettes and the other for wrapping materials as described previously.

9.8.2.1 Innovative Solution Adopted for the Wrapping Function: The Arbor

The key element of the packaging process in the ALFA machine is the so called *arbor* i.e. a square section steel pipe use as a mandrel to wrap the protective layers around it.

The functional cycle is rather simple and innovative:

1. The empty arbor is completely wrapped by the foil inner layer.
2. The packs end foil is closed and the bottom is pressed to obtain a flat, almost, sealant wrap.
3. The paper outer level is pre-glued and tight wrapped around the arbor leaving the upper entrance open.
4. Cigarettes are inserted in the arbor.
5. The pack is extracted from the arbor by pushing the cigarettes.
6. At the end of this sequence, an open pack is obtained.

The advantages of this process are mainly:

- Good quality of the cigarettes that are protected by the arbor.
- Excellent quality of the pack that has a squared shape corresponding to the arbor shape.

Keeping in mind the company know-how, it was decided to position the stations in the following way:

- First cigarette line: along a timing belt with containers.
- Cigarette compression: on a 4 stage indexed wheel.
- Arbor folding with foil and label: on an 8 station indexed wheel, called the main wheel.
- Pack closing: in an 8 stage wheel.
- Stamp insertion and pack exit: along another, second, timing belt with container.

The design was completed with the foil and paper line to feed material up to the main wheel.

The principal drawback is the very poor flexibility of the machine to accept any pack size change: to change pack size one has to substitute all the arbor and all ancillary folding units. The description of the machine is made starting from the wrapping material flow and then looking at the cigarette and pack flow. At the end of the design activity, a complete functional-logical documentation was prepared, Fig. 9.12.

Note: Paper and aluminum sheet feeding systems were almost identical for most of their components and have very similar functionality. The main difference is the end part of the path where only the paper is smeared with glue and fed in a vertical position while the foil is fed in a horizontal position. Here, after the paper system has been described in detail, and at the end of the chapter, some information is given on the aluminum foil feeding system.

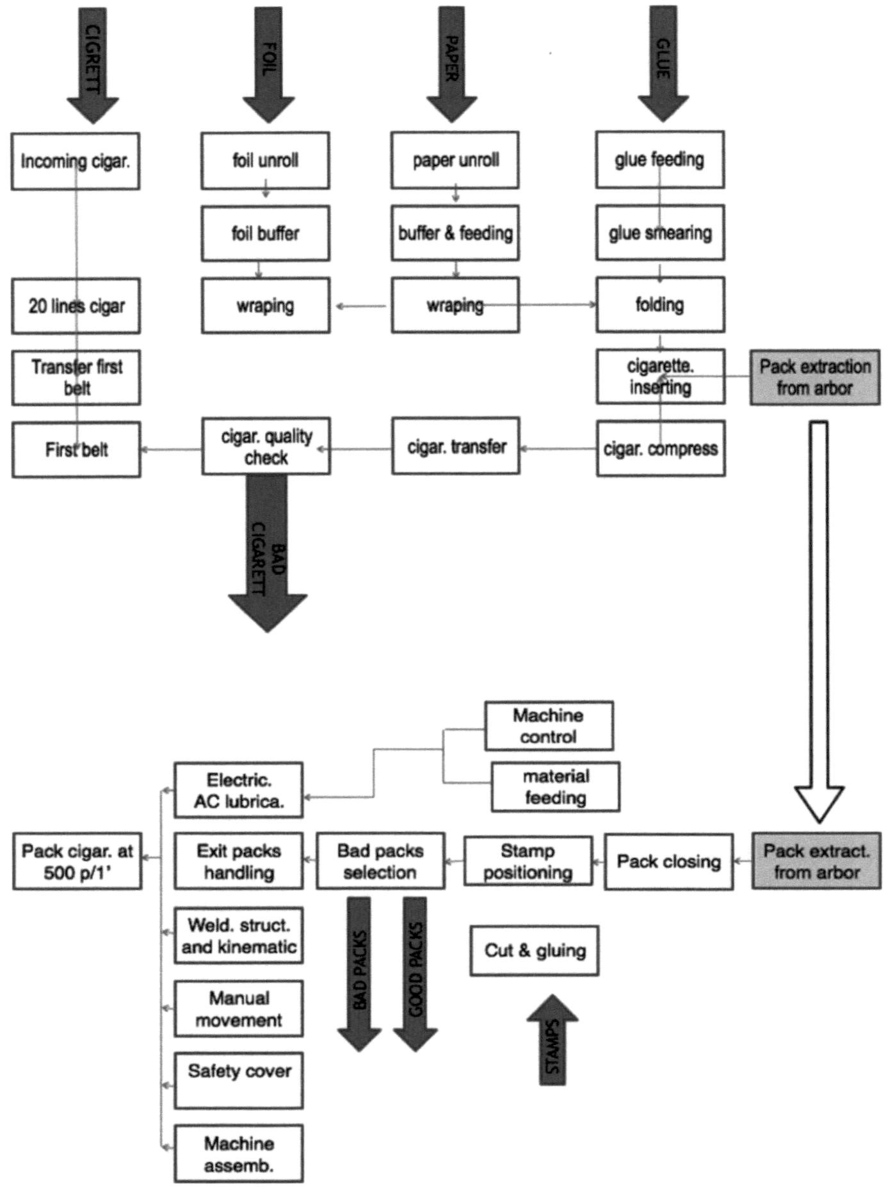

Fig. 9.12 Operational sequence: from left to right in the top figure and from right to left in the bottom figure

9.8.2.2 Solution Adopted for Cigarettes Line Sub-functions

1. Extract[13] cigarettes from the mass chute.
2. Place them in three layers with 7-6-7 cigarettes (2).
3. Check the number of cigarettes (3).
4. Check the quality of the cigarette tips (3).
5. Eject bad quality cigarettes if one is missing (3).
6. Compress cigarettes (4).
7. Push cigarettes inside the already wrapped arbor (1).

9.8.2.3 Solution Adopted for Material Line Sub-functions

1. Wrap foil around the arbor (1).
2. Press pack bottom.
3. Wrap pre-glued paper label.
4. Close pack bottom.

The two lines merge and then:

1. Extract cigarettes and wrapped envelope from the arbor.
2. Close the pack upper end.
3. Place the pre-glued stamp.
4. Extract good and faulty packs from the machine.

Some idle stations were placed along the cigarettes lines before the ejection station in order to obtain the "no bad pack" production functionality.

In fact, with this logical layout it was possible:

- To stop the material feeding in the case of missing or faulty cigarettes.
- To eject cigarettes and thus prevent the production of "bad" packs in the case of missing foils or labels.

> The last two functions were totally new on the market and a very strong selling argument. Both important functions were obtained, because it was possible to stop materials feeding at any time at full speed.
>
> Another change from the logical layout to the physical one was to insert idle stations in the final line with the goal of allowing the glue to solidify properly. In comparison with competitors and SASIB's previous products, this final line was rather short in relation with the short setting time of the hot melt glue selected.

[13]Functions described by imperative tense.

9.8.3 Function of Feeding Paper and Aluminum Foil

The function of feeding the two materials, paper and aluminum foil, is one of the most important functions to be performed by the packing machine. The feeding system function goes from picking up the incoming material, to positioning the cut and glued material in the right position, where the folding tasks start.

This function is very important for different reasons such as:

- High impact on pack quality: the exact position of the proper gluing is essential to produce a high quality pack.
- High impact on real quantity of packs produced per hour: the machine must run, if possible, at top speed during row material feeding by operators.
- High impact on jamming possibility: a large part of machine stoppage is caused by jamming in the paper feeding system.
- High impact on the ergonomic aspect because material feeding is one of the most time consuming activities of the operator in charge of the machine.

9.8.4 Solution Adopted for Feeding Paper and Aluminum Foil

The label feeding system was produced by competitors and used in old SASIB machines in two different ways:

- Pre-cut labels to be fed manually as a stack of few hundred units.
- Ribbon of labels fed as a reel. The ribbon must have a series of notches to be used by the traditional mechanical feeding system.

The design activity required a series of decisions that may be divided into two groups:

1. Design philosophy decision about the technology to be used.
2. Functional decision about the material flow handling.

The two decisions had to be revised repeatedly in order to optimize the material handling with the most appropriate technology: because of the circular reasoning it is difficult to define what decision was taken first and how the innovative layout was defined but we may record the decisions about:

Technology: the solution principle was to move the paper with an electric motor without any mechanical connection with the main machine system.

Material handling: the solution principle was to use a "Y" layout with two reels feeding a single final line; it was also decided to introduce a new functionality as capability of the system, so as:

- To switch from one bobbin to the other at full speed.
- To stop and restart feeding at top speed: i.e. the capability to "miss a label", starts and stops feeding on command without warning.

The paper feeding system was thus designed as a truly "independent" module connected to the main machine through electronic commands only.

> It is difficult to compare this functionality with the competitors' system at that time but we may say that probably none of the features were present on any competitor product.

On the basis of these requirements the subsequent design decision was:

- To use a non notched paper ribbon but marked with a timing spot.
- To design a stepper motor driving system capable of metering the ribbon.
- To design a cutting unit to generate the label.
- To use hot melt glue.
- To design a system to put glue on the label and move the label to its final position system, where the label matches the arbor.

Gluing a label at high speed with the required quantity of glue in a defined position is one of the "pitfalls" of any packaging system and, to simplify the system it was decided to use nozzle spray gluing system instead of the traditional smearing wheel system. In a few words, the functionality of the feeding system was:

- At the start to have motion as continuous as possible to unroll the coil smoothly.
- At the end to have intermittent motion to match the index movement of the machine.

To keep the ribbon tension as uniform and controlled as possible, upstream of the metering unit it was decided to install DC motors on the reel mandrel to unroll the ribbon with a constant tension independent of reel speed and, in particular, acceleration and deceleration independently of reel depletion level.

9.9 Experiments on Demonstrator Units

After these decisions, the "conceptual" stage of the design was practically finished and all major functions were defined.

Before starting the system embodiment design, "two demonstrator units" were built:

- A stand alone unit capable of feeding pre-cut labels from a reel.
- A nozzle spraying units to be installed on an existing machine running at 200 packs/min.

9.9.1 Gluing Function

The selection of the glue producer was started under the following requirements and constraints:

- To be accepted by the USA regulatory Agency FDA (Food and Drug Administration) as an "eatable" item.
- Not to release any smell or any potential cigarette contaminant both during production and during storage.
- To have a short closed time.
- To have low application temperature.
- To stand test conditions for high temperature and moisture resistance.
- And, last but not least, to be accepted by the main potential customer.

After some attempts and a series of laboratory tests (using an old packer at 300 packs/min), we selected two types of glue produced by two different companies and we got "the green light" from our major potential customers.

Both systems ran very well, no objection was made, the top management approved the solution and *the embodiment and details design* started.

It was only later that the spray gluing system showed up as impossible at top speed! The first machine prototype was equipped with electric, coil activated, nozzles to spray the glue but the very first test showed that the dots of glue applied on the label were unable to withstand the acceleration produced by the folding. In fact, the glue was "centrifuged" from the label and then covered a large part of the machine with a thick layer of sticky glue: a real mess!

It was thus necessary to go back to the traditional smearing wheel system, losing 6 months of time.

Why did this error occur? The test were not severe enough because they were done at a slow speed (300 packs/min), compared with the machine target speed.

Lesson learned: feasibility tests are key factors and must be run in the worst possible condition. At that time there were no high speed packers available for testing and the low speed performance was not critically examined.

We will not describe the unsuccessful glue spreading and we will describe the smearing wheel version installed on the packer.

The use of a stepping motor and marks on the paper made it possible remove the notch used in all machines to advance the paper: this was a strong point but it was heavily criticized because we lost the capability to use standard "notched" ribbons: another battle against conservative minded people, won through the support of the really "friendly customers"! Speaking about potential customers, from the very start of the project we selected Philip Morris of the Richmond VA plant, as a "friendly customer" and we cooperated with them for the entire development time, up to prototype testing, in a real industrial production environment.

Two Stories About Reliability and Dynamics of Human Behavior

Speaking about the motorized mandrel, it may be amusing to mention two stories that show what *product development* activity really means.

The mandrel where the incoming bobbin is placed has three radially protruding teeth, pushed out by compressed air to secure the coil rotation. One day the shaft was tested, without the paper coil, on a bench to measure torque and speed when, all at once, the motor went into a "full speed ahead" condition and the teeth, without the bobbin usually holding them in position, were ejected: the three parts, 50 gr. each, took different directions: one went up, hitting the room ceiling at 5 m height, another stopped against the floor while the third crossed the testing room, hit a glass window and went through it and finally stopped against a wall. Incredibly, no one was hurt. The investigation showed a long series of partially connected causes and remote causes and reinforced the idea that *testing and lab activities are very dangerous*.

Another short story may well illustrate the intricate process of design of a mecatronic element and the constraints of some decision.

At a certain stage of development, a new manager was willing to "check" the design of the ALFA. A consultant was hired to check if all "commercially available" components were properly sized and selected. The main result of this investigation was that the DC motors to unwind the coil were considerably oversized.

The investigation showed that, at the time, the first prototype utilized the same DC motors directly available without buying new ones: they were large enough, where assembled, tested, approved and inserted in the BOM without any further refined computation: who was at fault?: the designer, the BOM handler, the head of prototype construction or the director of R&D? We do not know the correct answer but we must consider whether it is appropriate or not to reduce the number of parts in stock and their variety [1].

(BOM is an acronym for a bill of materials or product structure. It is a list of the raw materials, sub-assemblies, intermediate assemblies, sub-components, parts and the quantities of each one needed to manufacture an end product).

References

1. JMAC (2018) Variety reduction program. https://english.jmac.co.jp/wp-content/uploads/2010/07/rd20f.pdf
2. Techreport (1976) Robotica e Automatismi Multiscopo per Industrie Manufatturiere. Technical report, Olivetti spa
3. Ulrich KT, Eppinger S (2008) Product design and development, 5th edn. Mc-Graw Hill. ISBN: 978-0073404776

Chapter 10
Embodiment Design of the Packaging Machine: Prototype Development

Abstract This chapter is devoted essentially to the embodiment design of the SASIB ALFA Packer Prototype, following the topic dealt with in the previous chapter. The embodiment design is the part of the design process in which, starting from the main decisions made in conceptual design, the product plan is concretely developed, dealing with technical and economic criteria. This step moves from logical to physical solutions, to define shape and size. In the example shown, the embodiment design is essentially an illustration of the main constructive solutions adopted. However, unlike conceptual design, embodiment design requires many corrective actions in which analysis and synthesis alternate. It is a less systematic process that gives rise to many reviews. In any case, we do not deal with the detailed design phase.

10.1 Subdivision in Modules of the Machine Front Side

From here onward, the design requires decomposition of the system into key-modules with a good level of independence, such to allow separate development [2]. We may start the description of the different modules that compose the machine, with the front view, Fig. 10.1.

10.1.1 Modulus of Paper Feeding System

After the system requirements and system conceptual design, let us describe the paper feeding system, Fig. 9.11 in Chap. 9.

10.1.1.1 Paper Roll Support

The paper roll is blocked on an expansible shaft connected with a DC motor torque-controlled. The axial position of the roll is actively controlled by a stepper motor on each shaft action by a pair of edge-controlling photocells thus allowing the use bad bobbins with conical shape.

A. Freddi and M. Salmon, *Design Principles and Methodologies*, Springer Tracts in Mechanical Engineering, https://doi.org/10.1007/978-3-319-95342-7_10

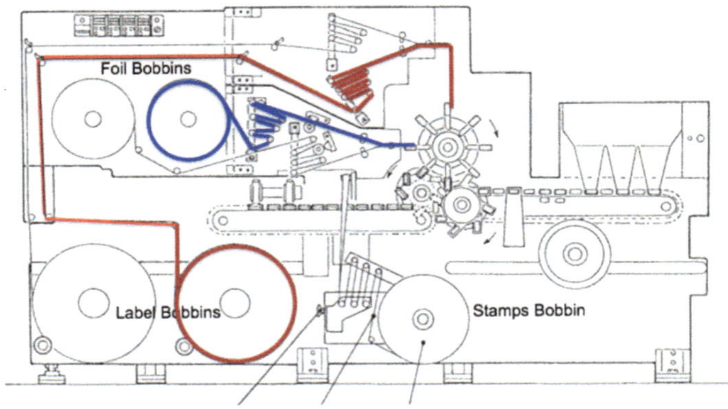

Fig. 10.1 General front view of the machine

10.1.1.2 Swinging Tensioning Arms

The paper leaving the roll with an almost continuous motion is threaded through an oscillating arm with three rolls matching another three fixed rolls, Fig. 10.2. The design of this very traditional device shows different critical parameters. The oscillating arm should have contrasting properties:

- Large buffer of material to allow the start of the abrupt unroll movement at top speed in the case of coil start up from an idle roll, when there is a switching from one coil to the other.
- Small buffer to allow gentle slowing down of the coil in case of feeding line stoppage, see also Fig. 9.11.

To satisfy both cases a very large arm stroke should be necessary, with a too large inertia. The force made by the downstream pulling device is proportional to the inertia reflected by the arm and the rolls on the ribbon. To solve the problem two different decisions are needed: a mechanical one and a control one, as a good "system design" requires.

1. Mechanical solution: all rolls are made in light resin with special roll bearings to minimize inertia value.
2. Control solution: the oscillating arm is forced to operate in two different central positions:
    - Storing a minimum quantity of paper when operating at steady state high speed to be ready for any stoppage, keeping the position when the line is in operation.
    - Storing a maximum quantitative of paper to be ready to start from stoppage to full speed, maintaining position when the line is stopped.

Fig. 10.2 The swinging arms of the two lines for paper and the two lines for foil feeding

Downstream of these "tensioning arms" the two ribbons converge in a small, compact innovative unit to emerge as a single flow of cut foil as shown in Fig. 10.3. The unit functions are:

- To pull and meter the required length of paper from the dancing arms upstream, from one of the lines.
- To keep the other ribbon in stop position, ready to be fed any time.
- To cut the paper to the required length and leave it to be pulled downstream by the subsequent gluing device.

The metering functions consists of two pairs of pinch rollers directly coupled with stepping motors programmed to cut in phase with the printing register.

The cutting system is also rather innovative in design and consists of a rotating shaft, with around 30° movement, to generate a scissor type cut: this solution has the possibility to make a non straight cut and to operate like scissors, reducing the torque requirement for the stepper motor. Stepper motor functionality and timing will be described later.

10.1.1.3 Label Down-Feed Units

This unit picks up the foil in horizontal situation, bends it down, glues and "shoots" it into the position to match the arbor, i.e. the metal box that it must be wrapped around. It is one of the most important functions of the machine:

- Proper gluing is fundamental to get a strong pack but any spread of glue will quickly stop the machine.

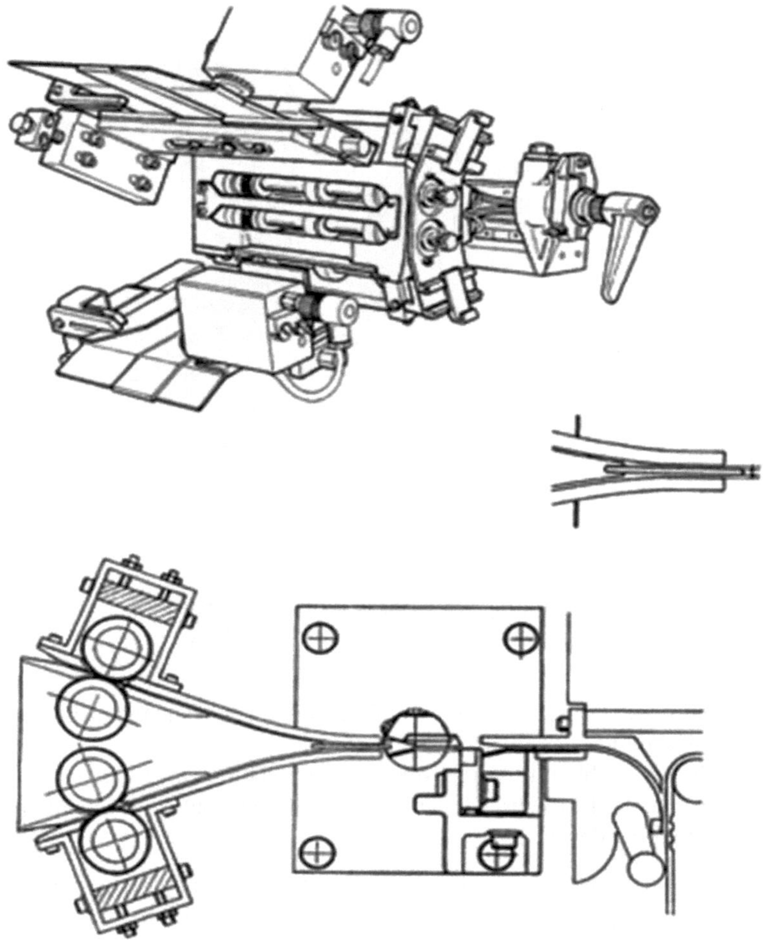

Fig. 10.3 The core of the paper feeding system: the metering and merging unit where the two ribbons are metered by two independent stepping motors

- Paper position of the label in its final position is the only way to get a good pack with the right position of artwork, without skewed sides. These parts have many critical design features because any friction of the paper may interfere with its position and because proper gluing deposition is hard to obtain.

The unit has two main functions:

- To smear glue with a smearing wheel that picks up hot glue from a constant level small basin.
- To move the foil down, from the position when leaving the cutter to its final position.

The smearing wheel is the traditional type and is moved with an indexed motion by a single stepper motor while the label is moved by two small timing belts by means of another stepper.

The independent drive of the smearing wheel is an important innovation, allowing the feeding of "non glued" paper for testing.

- Because glue deposition is to be avoided as much as possible, the unit is "lightened" to reduce the area of possible glue deposition. For example, the label is kept in position by wires that operate against the timing belts.
- All the units may be easily opened to take out jammed paper and to clean.
- Downstream of the belts the foil is "dropped down" toward it final position.
- The unit is completely computer controlled and all timing, strokes and dwell time are easily programmed by the end-user.

> At that time, the competitors' paper feeding systems had no paper feed and no start stop at full speed and no software control. This is one of the most "clever" devices of the packer and one of its strong points.

Performance

In terms of dynamic behaviour the main data are:

- Feeding lengths 160 mm
- Feeding time 23 ms (corresponding to the time that the arbor is standing and its clamps are open)
- Average speed 70 m/s
- Top speed 140 m/s
- Acceleration 1000 m/s^2 (=100 $\times g$)

The oscillation period of the cutting roll is a few (a dozen) milliseconds but this data is software defined.

10.1.1.4 Control System of the Paper Feeding System

All the functions of the paper feeding system are controlled by a single board computer produced by the Intel Co. The control system is connected with:

- The machine level control with very simple parallel interface for input and output data.
- The push button human interface.

Information exchanged with the machine level control are:

- A trigger signal to synchronize paper feeding with the machine cycle. This signal is, as we will see later, generated directly by an encoder" to avoid any delay.

- A parallel coded "status" signal with information on the "operating mode" such as: dry running, running, "miss label command". With this solution it is possible to skip a foil in the case of missing cigarettes and to perform the function "zero waste" (i.e. no production of faulty or cigarette missing packs).

The single board computer gives back a signal to define its condition: ready to run, one coil near to empty, jamming. All these conditions required stopping or slowdown. This information is also transmitted to the CRT colour human interface.

It is also possible to manually control similar functions such as:

- Oscillating arm positioning for paper thread.
- Switching from one bobbin to the other.
- Feeding single foil manually without glue.
- Glue smearing wheel rotation.

10.1.1.5 Aluminum Foil Feeding System

The produced pack is made up of an internal layer of so called "aluminum foil" fed on a coil to the machine. This system is identical to the paper one but missing, obviously, the final part that glues the paper. The line ends with a cut foil in the proper position to be picked up by the "arbor" as described later.

All the components (from bobbin shaft down to the metering-cutting device and the control system) are exactly the same as the paper line previously described.

10.1.2 Modulus for Cigarette Feeding and First Belt

Cigarettes, entering with a direct link with the "maker" machine, are fed to the packer with a sort of vertical chimney full of parallel cigarettes falling down in a mass flow. The chimney feeds a sort of inverted funnel to open the flow and let the cigarette enter 20 single cigarette channels. The cigarettes' entrance in the channels is eased by 21 alternate rotating cylinders.

At the end of each of the 20 channels the cigarettes are pushed, by an alternate motion slider, moved by cams, into a series of 20 cigarette boxes installed on a timing belt with indexed movement, Fig. 10.4.

The transfer is made in three successive stations feeding 6, then 7, and finally 6 cigarettes to fill a box installed on the first belt. This transfer is one of the most critical actions for the cigarette quality because cigarettes are made of less than 1 g of tobacco strands and are easily damaged by any manipulation or strong acceleration: in particular, transfer from the chutes to the box must be done with a rather gentle movement to prevent tobacco "sliding" inside the cigarette or the paper end hitting a fixed wall or the formation of the so called "white collar" a very bad quality mark.

The design of the law of motion of this plunger is very critical and it was redefined many times, increasing to the maximum the stress during the return stroke, to allow

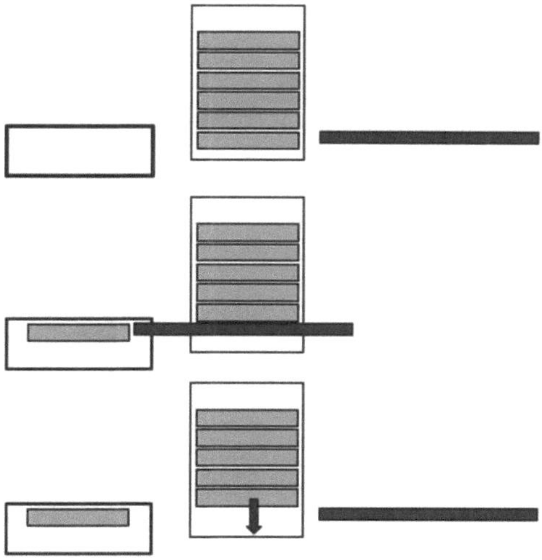

Fig. 10.4 Transfer of cigarettes from the hopper to the boxes placed on the first belt

Fig. 10.5 The station to transfer cigarettes to the box on the first belt

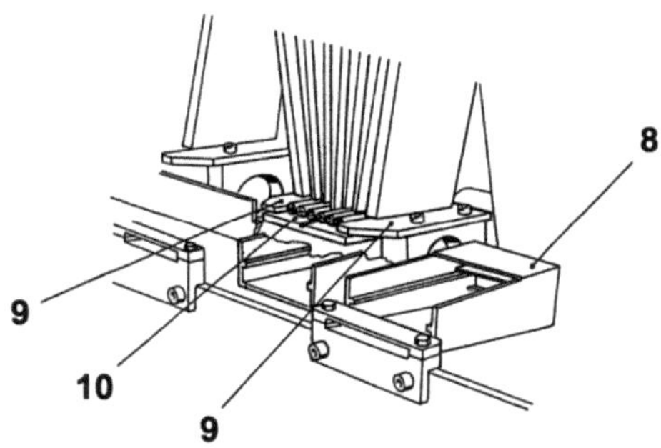

more time for the onward stroke responsible for the cigarettes quality, Fig. 10.13, at the bottom. But the cam redesign is not enough to achieve a good quality, so a "slow down" receiving device is inserted in the machine.

To reduce cigarette axial acceleration and to prevent the cigarettes being stopped against a fixed wall, the transferring system is completed with a moving counter-plunger, having the task of gently slowing down the cigarettes. This sort of receiving device is moved by a specific pair of desmodromic cams, Fig. 10.6.

The law of motion of this part is mathematically defined by a seven order polynomial obtained to respect the desired constraints for the movement:

- A plunger pushes a pack or a cigarette with its law of motion.
- During the deceleration phase the cigarette, represented in Fig. 10.7 with a straight line, continues its motion at a constant speed. In this case the cigarette would hit the limiting motion wall at top speed.

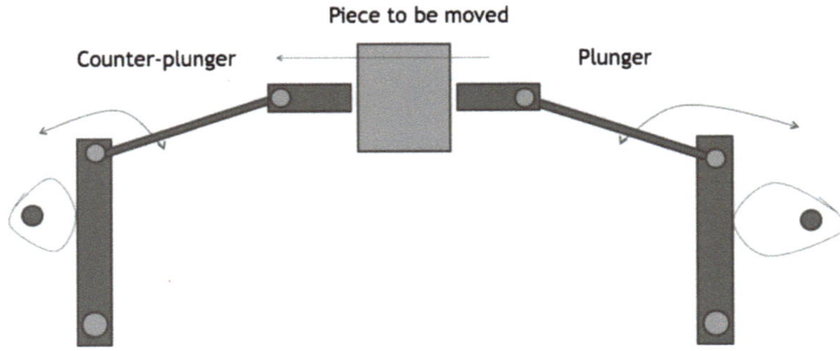

Fig. 10.6 The combined kinematics of the plunger and counter-plunger to obtain gentle movements and stopping of the cigarettes pack

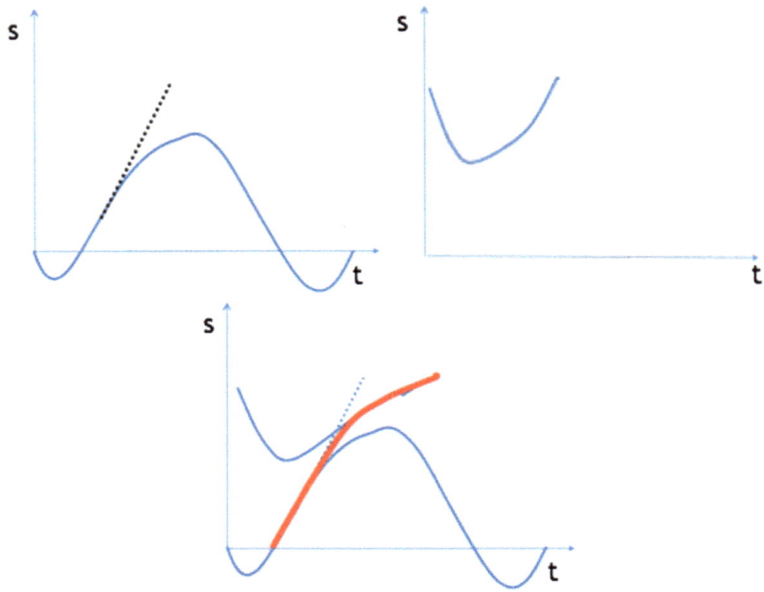

Fig. 10.7 Laws of motion of the plunger (on the left), law of motion of the counter-plunger (on the right). Composite motion of the cigarette (bottom). The dotted line represents the motion of the cigarette without the counter plunger

- The counter plunger positioned opposite the plunger will move with its own law of motion.
- The counter plunger will be hit by the cigarette and will slow down the motion gently.

The "box" containing the cigarettes properly positioned is moved by the timing belt through a "control station" where 40 gentle pins are moved against both sides of the cigarette to check for cigarette presence and for good tobacco side tip quality. The pins are gently spring-pushed to guarantee not only the presence of the cigarette but also the compact aspect of tobacco tips.

When any cigarette is faulty, the whole bunch of 20 cigarette is blown out by an air compressor blow. The machine is not stopped but all material, foil and paper flows are interrupted to avoid producing an empty pack or generating a paper jam.

10.1.3 Modulus of Compression Drum and Arbor Drum

The cigarettes positioned in the arbor are slightly compressed, just as they are in the finished pack sold on the market. Compression must be strong enough not to lose cigarettes but gentle enough not to damage them: the right compression is found by experience and depends on the so called *firmness* of the cigarettes i.e. the elasticity of a cigarette under radial load.

Transfer of the product from the loose status on the first belt to the "gentle" compression in the arbor wheel is provided by a "compression turret". It is made by a series of boxes, or pockets of variable dimensions.

The process sequence is the following:

- Opening of an empty pocket to allow transfer from first belt.
- Rotating of the wheel and contemporary gradual gentle compression of the pocket on the cigarette bundle.
- When they arrive in front of the arbor, the pockets must be opened and allow transfer of the cigarettes into the arbor.

The cigarette compression[1] must be gentle so as not to damage the roundness of their shape and the opening must be fast enough to leave the cigarettes bundle still compressed and so let them enter the arbor without high friction.

Moreover, the mechanical operation of the compression drum is also critical because the two halves of the pocket are guided by a slide guide and there is little room available for the opening spring, Fig. 10.9. The same Figure also shows the arbor drum with its eight stations (A, B, …, H).

> In the original design, the compression drum was made with four stations and the opening of the pockets was very critical. Over a certain speed the springs were not strong enough to open the pocket and the plungers were not able to push out the cigarettes: the results was that the pocket remained full of cigarettes and, on arrival of the subsequent cigarette batch, all the cigarettes were pushed away with random stopping of the machine and, sometimes, with damage of the plunger. It took great effort to design special titanium springs small enough to stay in their housing and strong enough to overcome the friction of the slide guide.

[1]The process is a sort of "black magic" because the relaxation of tobacco is exploited to give the cigarettes bundle to enter the arbor lightly compressed.

The complete solution of the problem was found only some years later when the ALFA design was reviewed, the turret was modified from four stations to eight stations, as it is today, and shown here.

These spring related problems are counterproof of the fact that the utilization of springs always gives rise to functional problems.

10.1.4 Modulus of Arbor Drum

Here is described the core of the packer[2] where the three main components of the pack, cigarette foil and paper, mate and the pack leaves the wheel completed with just only one operation to complete the cycle, Fig. 10.9.

The functionality of the wheel is very critical because all the actions are performed on paper and foil and are difficult, if not impossible, to "model" or even to test: all bending, folding, joining must be designed on the basis of trials and errors; moreover, it is difficult to evaluate the quality obtained that is based on visual inspection only.

All operation is also sensitive to speed in an unpredictable way, probably with a quadratic effect: some operations may be successful up to a certain speed and become catastrophic over a certain threshold.

To have an idea of those problem one may refer to sheet metal bending and forming where, in spite of dealing with a well defined material, all dies must be "biased" in order to get the size you want: but here we must deal with paper, always undefined and with changing properties depending on actual moisture content, temperature, aging, producer, batch and even original transverse position on the paper mill.

Another difficulty arises from the effect of material quality on the results: paper weight, rigidity, spring back, "wettability", have a great and immeasurable effect on the final quality of the pack.

The wheel is, moreover, the location where much of the jamming occurs and thus has a very strong effect on the real productivity of the machine and its runnability.

We have to say that, here, we were lucky and not only was the design "straightaway": merit of the combination of "senior draftsmen" with long experience and "newcomers" taking care of design ingenuity and innovation capability. The real field experience at customer factories did show a very good flexibility and adaptability of the ALFA packer to deal with almost any

[2]The problem is: how long and with what forces we have to act to guarantee not to have any spring back?

material in any place from USA air conditioned plants to rather rough Asian factories.

Taking into consideration this condition it was decided to:

- "Mimic" as far as possible the action performed by the previous SASIB packer.
- To allow easy access by the operator taking away any obstacle in front of the wheel.

10.1.5 The Wrapping Sequence

The core of the stations is the arbor and its two clamps that keep the folding material in place during the wheel rotation.

To better understand the arbor drum operation we describe here the sequence of operation to get an almost complete pack, Figs. 10.8 and 10.10:

- STATION A one clamp is opened to allow the foil insertion, the foil is then folded during the wheel rotation by two fixed folders.
- STATION B the foil is folded completely around the arbor.
- STATION C the paper drops down vertically and is clamped and then folded during the rotation.
- STATION D the paper is wrapped completely around the arbor.
- STATION E the back of the pack is closed by three arms.
- STATION F the back of the pack is compressed by a plunger to get a flat surface.
- STATION G the cigarettes are inserted in the arbor.
- STATION H the pack is transferred to the second belt and the arbor is now empty and naked.

The sequence is the real core of the entire machine and the source of product quality: any small misalignment of the paper or foil would give a faulty pack or, in the worst case, a loss of material and then a jamming condition.

At the last station, the pack full of cigarettes is transferred to the third wheel completely finished but with the top part still opened.

10.1.6 Modules of Last Front Part of the Machine

The remaining part of the machine performs the rather easy task of closing the pack and applying the stamp:

- A third wheel.
- A final toothed belt with containers.

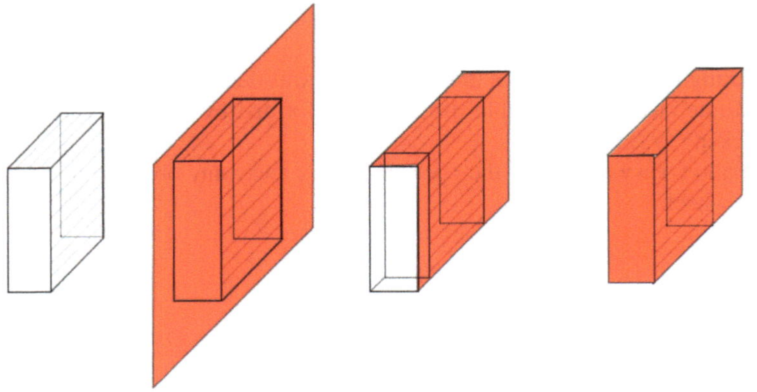

(**a**) *Naked arbor*

(**b**) *Paper is cut and glued aside of the arbor*

(**c**) *Paper is wrapped around the arbor*

(**d**) *The bottom is closed*

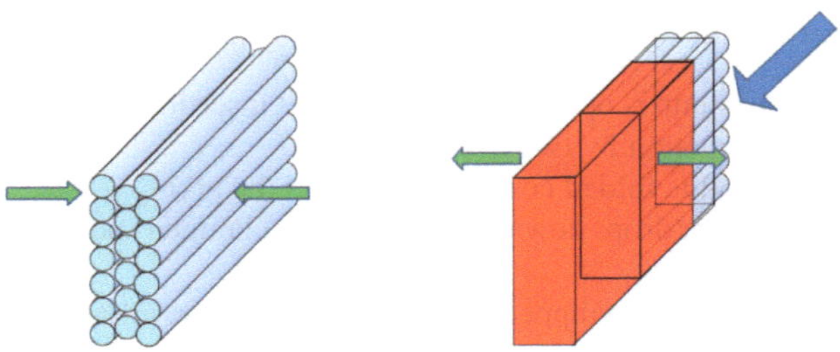

(**e**) *Cigarettes are compressed in the compressor drum*

(**f**) *Cigarettes are pushed inside the arbor where they meet the arbor walls with a slight relaxation*

Fig. 10.8 The packing sequence of all SASIB machines: the cigarettes are protected by the arbor to guarantee their high quality

- Exit belts to move the pack out of the machine toward the over-wrapping machine: one for good packs and the other for packs without the closing stamp.

10.1.6.1 Pack Closing Wheel

The next function of the next and last wheels is to close the pack upper end. The wheel is very simple with four stations:

- The first one to receive the pack from the second wheel.
- The second to fold the short side of the pack.

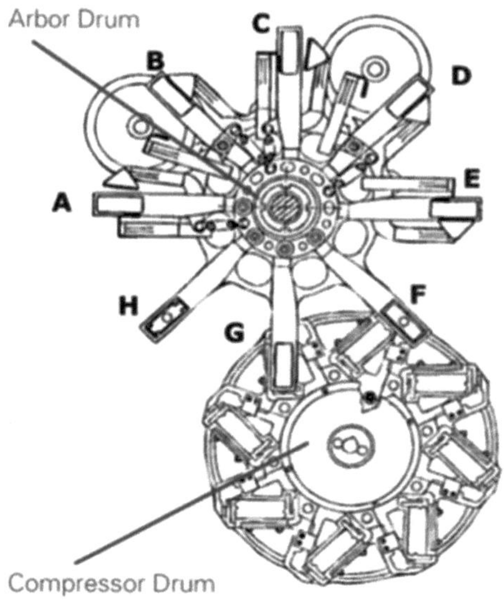

Fig. 10.9 Modulus of the compressor drum together with arbor drum

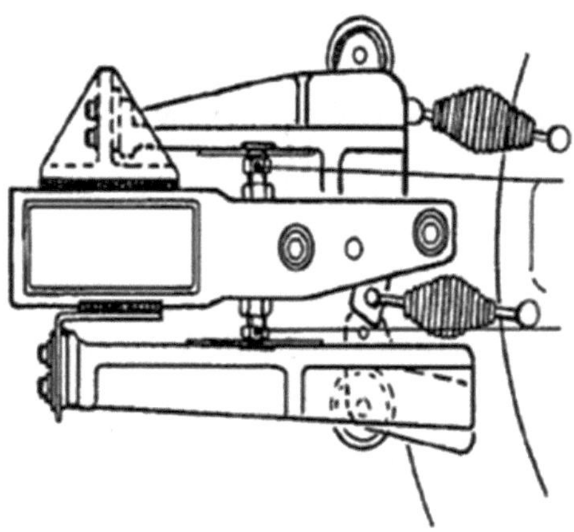

Fig. 10.10 Detail of the arbor with the two levers that keep foil and paper in position and that open at certain stations

- The third to fold one of the long sides and then, during the movement to close the other fold.
- At the end to transfer the pack to the successive "final tooth belt".

10.1.6.2 Final Tooth Belt

The so called second or final belt looks like the first one: it consists of a timing belt with an indexed movements and series of resin boxes to contain the finished packs. The functions performed by this belt are:

- Holding the packs firmly to allow the hot melt glue to cool down and dry.
- Placing the so called fiscal stamp to close the upper part of the pack and leave time for its glue drying.
- Eject packs that are faulty, for any reason, in a separate exit line.
- Deliver packs on a transfer belt, or chute or input system of the downstream cellophane over-wrapping machine.

The length of the belt is made up of a dozen of stations in order to achieve a cooling time of about 15 s, much shorter than the drying time of traditional cold glue.

The boxes on the belt are not so critical as the first one because the design is much simpler, tolerance are larger and they have no direct contact of plastic material with cigarettes.

10.2 Subdivision in Modules of the Rear Part of the Machine

We have seen the description of the front part of the machine and now we are going to describe the "mechanical drives" of the ALFA i.e. the part where all the motions are generated, and the active components are supported. The mechanical part of packaging machinery is "conceptually design driven" by the front part and all units requirements are defined by the functionality of the relevant active parts placed at the front. Design requirements, such as speed, law of motion, acceleration and jerk limitation are really "born" in the active part of the machine and transferred, as required, to the mechanical side. The design is less "creative" and is a sequence of "problems to be solved" often not trivial, because of high speed and compelling requirements.

Figure 10.11 gives an overall picture of the phasing of some of the machine components: 36 moving parts to be phased with very few degrees of allowable error.

The design of any packaging machinery is driven by the process that must be carried out: this is a very specific peculiarity of all machines that perform "handling" operations and not "technological" operations.

Before describing the machine, we may have a look at the status of the art in this sector at the end of 70s.

Generally speaking, we have to remember that at the time of the ALFA design start up the PC was not yet on the market and computations were made mainly with a remote computer under the GE service in "time-sharing" with teletype as a terminal on a commuted telephone line, with transmission speed of few hundred byte/s! At the ALFA start up time the design office was using hand cranked computers and there were no electronics for operations calculators! The first real computer for technical application bought by SASIB was a Digital Vax in around 1982.

Up to 1982, about all computations were made through the University (in this case Bologna University, Applied Mechanics Department).

To give an idea of the design methodology at that time, we may remember that to evaluate the axial inertial value of any components the Huygens-Steiner theorem was experimentally used: today any CAD performs this calculation in fractions of a second!

In Bologna area there was considerable experience on design of intermittent driven machinery and a certain familiarity with index motion driving systems and cams. Sweets over-wrapping machines were running at 600 packs/min.

The design office consisted of designers with long experience but not university graduates: at SASIB in 1979 there was only one graduated engineer capable of performing some computations among about 40 mechanical designers or draftsmen, although, very complex problems were solved on the basis of intuition and inventions of very skilled personnel. Nevertheless, the lack of computers but considerable "hands-on experience" kept the application of theoretical knowledge at a minimum level.

Looking back at the gained experience, the mechanical part of the machine was a good design case with very few mechanical failures, while noise and vibrations were at acceptable values. The only mechanical failure to be recorded was originated by the cone-type shaft connector and excessive shaft bending deformations.

We may now divide the mechanical drives into a number of "conceptual" sub-assemblies:

- Indexed moving parts of the three wheels.
- The four timing indexed belts.
- The seven plungers transferring the product from the first belt to the second belt.
- The continuous motion shafts and gears to properly connect all the machine moving parts.

It is important to note that, different to all machines under production at that time, the ALFA had many functional sub-assemblies with only a software timing connection: The two web feeding lines had their separate motors without any mechanical link with the main mechanical unit: we may say that the ALFA was one of the first "distributed motor" units, a solution almost non-existent at that time and today almost a standard for any packaging system.

10.2.1 Mechanical Components

All the mechanical moving parts are connected with a single degree of freedom with the movement made by the central main motor with AC variable frequency (only one in the first version).

1	Main Motor		
2	Main shaft		Transmission to 2 and 3
3	Encoder		
4	Shaft	Cigarettes and pack transfers	Transmission to 5
5	Detector Shaft		Transmission to 6 Coupled with 7
6	Shaft	Hopper Feed Plunger cam	Transmission to 8
7	Hand wheel	Manual operation	
8	Intermittor	First belt	
9	Shaft	Top cams	
10	Angle gear		Transmission to wrapper forming mechanism
11	Shaft	Harbor turret index cam	
12	Shaft	Foil bottom folder cam	
13	Shaft	Index cam for the pack top folder turret	
14	Shaft	Stamp applicator assembly	
15	Shaft	Index cam for second belt	
16	Shaft	Index cam for compressor turret	

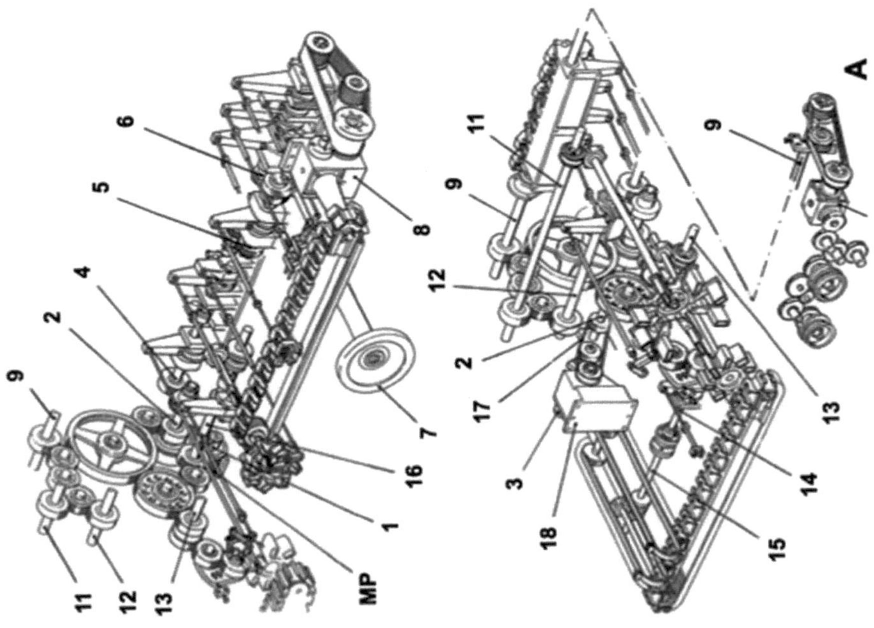

Fig. 10.11 View of the main mechanical units of the machine: the index cams and the series of desmodromic driven pushers

The main mechanical components of the machine may be described as follows, Fig. 10.11.

1. A series of shafts in a longitudinal direction of the machine.
2. Several desmodromic cams rotating on these shafts.
3. Four index worm type cams to give indexed motion to the two timing belts and three wheels.
4. A parallel "para-block" index cam to move the final exit belts.

This layout solution was, somehow, similar to the old SASIB machines where the "core" consisted of barrel, or cylindrical type cams, unlike the competitors that were using all parallel shafts put in a transverse direction and moving the indexed part with Malta cross units.

The two alternatives were not studied in detail, as the theory of new product development would have recommended, but the longitudinal shaft solution was selected mainly on the basis of a certain similarity with the old SASIB model.

All the longitudinal shafts were connected, in the original design, through timing belts driven by the main electric motor.

The main timing belt, moved by the main motor and moving many longitudinal shafts is, obviously, a key element: if the belt breaks, the plungers will cause a series of collisions that would practically destroy the machine.

The choice between a timing belt connection and a more traditional gear cascade connection is in favour of the belt, but after splitting the belt in two parallel belts and with a couple of switches to sense each half belt integrity: if one belt breaks, the other one will be able to keep the timing and slow down the machine to a stop. To increase safety, a disc brake was installed on the motor shaft. At machine testing, the belts caused a certain amount of noise and, later on, the machine was modified with the installation of gear cascade drive.

The main motor was an AC powered by a variable frequency drive to speed up the machine gradually; some measurements showed a large fluctuation of the actual speed during a machine cycle, caused by the index movements variable inertial force; for this reason the AC motor was, later on, substituted by a brushless torque motor.

Let us now describe the main mechanical components.

10.2.1.1 Index Cam

The old SASIB machine used immersion lubricated barrel, cylindrical cams with a milled groove but this solution was immediately discarded in favour of worm type grinded pre-loaded barrel cams because it was the only solution possible at the machine's target speed.

Because of the difficulties in machining such type of cams in the internal workshop, the solution was taken to buy them. All index cams were standard models produced by a Belgian company. No failures occurred with these parts.

There are five index cams to drive:

- The first belt.
- The compression turret.

- The arbor main wheel.
- The pack closing wheel.
- The second belt.

A "parablock" index cam, (i.e. a parallel shaft index unit) is used to drive the two final pack exit belts.

10.2.1.2 Desmodromic Cams

The transfer of cigarettes and packs from one wheel to the next one is made by a pusher with a stroke length of the order of magnitude of the cigarette. There are five main linear movements within the machine:

- A first group of three pushers, having the same movement, to transfer cigarettes onto the first belt.
- A single pusher for transferring cigarettes to the compression turret.
- A pusher to "compress" the pack bottom.
- A pusher to transfer packs out of the arbor wheel.
- A pusher to transfer packs onto the final second belt.

Moreover, three other pushers are necessary:

- To move the two cigarette quality control units.
- To generate a short stroke movement, a few millimeters, to align the cigarettes properly in the compression turret.

A further linear movement is used, as described above, to act as a "counter-plunger" to slow down gently the cigarettes in the first transfer from the hopper to the first belt.

The linear stroke alternate mechanism is typical of all packaging machinery and almost all machines are using "plate cam" with springs for the return motion.

One of the first decisions is to use desmodromic cams to avoid the use of springs and to define a standard design to be used in all components of the machine. The standard kinematic layout is characterize by an oscillating arm made of steel sheet bent and welded, with rolls. One of the two rolls was assembled with an eccentric shaft to allow for play control at assembly stage, Fig. 10.12.

Later on, the sheet arm was substituted by a cast iron one because of the poor resistance of the welding under variable stress.

The "good" handling of the cigarettes is one of the requirements and therefore the cam profile design is a very critical role.

The cam generated motion is made by an "active stroke" operating against the product and "return" stroke to be made as fast as possible in respect of stress on the cam and, therefore, two different laws of motion are foreseen. It was difficult to exactly define the requirement of both laws and, in fact, the cam profile was redefined many times during the machine development.

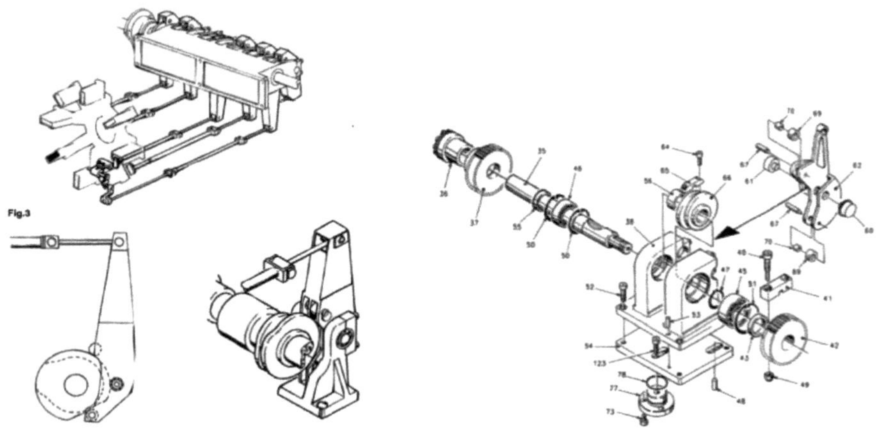

Fig. 10.12 A desmodromic cam drive of an oscillating arm: the shaft, the pair of cams, the oscillating arm, the steel saddle and relative bearings

In this activity, the collaboration with Bologna University was of fundamental help. The law of motion of the cams was really a "battle ground" with many sequential improvements and was, in large part, made with consultancy of the Bologna University [5].

The camshafts are pre-assembled on a saddle to speed assembly and to have good rigidity: overall, the cam system is very compact, reliable and with very good quality in terms of cigarettes handling, The cam for forward motion has many different stringent requirements:

- To keep products, cigarettes and packs as smooth as possible and to reduce to a minimum any "hit" or movement discontinuity.
- To maximize the active forward time with respect to the return phase.
- To limit Hertzian force against the cam.

10.2.1.3 Phases

Theoretically, machine movements coordination is very simple; at each machine cycle:

- Indexed parts move for 33% of the cycle time and remain stationary for 66%.
- All pushers move forward about 2/3 and backward for 1/3 of the cycle time.

The presence of an absolute encoder, (one turn each machine cycle), is very important, allowing an easy definition and setting of any movements. But real phasing is much more complex for many reasons as:

- The movement controlled by a step motor operates at constant speed to avoid resonance condition.
- Some movements had special conditions and must be phased with a given delay.

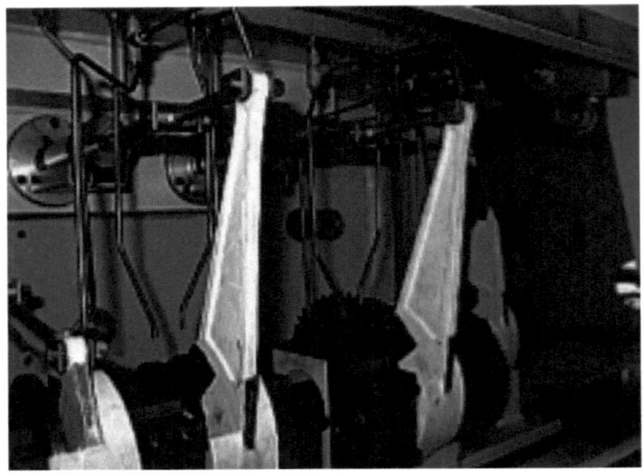

Fig. 10.13 View of the desmodromic cams from the machine rear. The levers have a different design and are in cast iron instead of bent sheets

We see in Fig. 9.9 that 36 moving parts must be phased with very few degrees of allowable error.

The original 1984 design had a rather fast timing without any care for optimization: later on (1986) the whole timing was redesigned taking care to reduce, as far as possible, synchronous movement of any long run stroke. The benefits in terms of structure vibration, noise, main motor torque behaviour were considerable: this is the most important improvement made at the machine design after the original design.

10.2.2 Stepper Motors

One of the first design decisions was the use of an independent motor to drive any part that had no risk of collision in the case of loss of position control because, at that time, electronics was still seen as a low reliability technology.

Therefore the core of the machine, the belts and index wheels, are driven by a single central motor but all feeding lines are driven by independent, electronically controlled motors: in the case of loss of control there is no risk of collision but only loss of functionality without any damage to the machine.

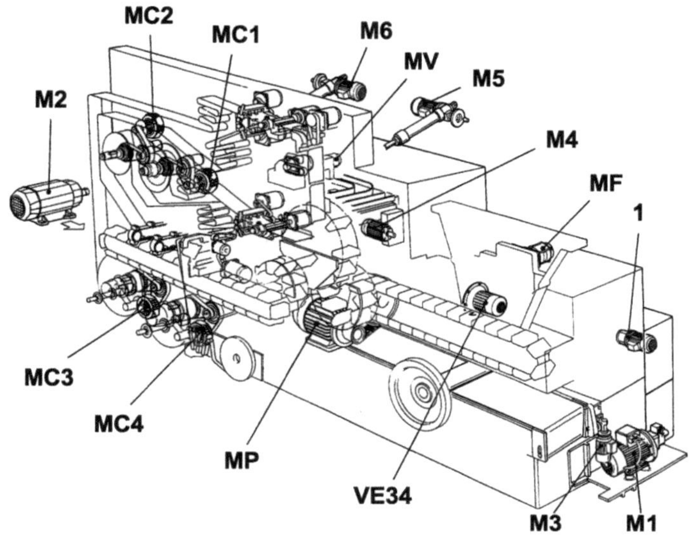

Fig. 10.14 The main features of the ALFA packer: general view of all the independent, computer controlled motors

The decision was further enforced by the impossibility to drive indexed parts with electrical motors: at that time, direct drive brushless motors were not yet available.

The second main decision about motors was to use open loop step motors instead of the more widely used CC close loop control motors. Here too, the Olivetti experience was fundamental: the stepper motors were used on the Olivetti SIGMA robot and were produced by MAE, an Olivetti fully owned company.

The reasons for this decision were:

- A technological reason, because the maximum allowable torque to inertia ratio is much higher for stepper motors than for CC motors.
- A managerial reason, because there was no experience available at SASIB on close loop control and the complexity of the CC plus encoder was judged too costly and complex.

Looking back, it was a wise decision: motor types have not been changed in the subsequent forty years.

The motion laws of all motors are characterized by:

- Short stroke: 200 mm of paper, few mm of stamp, one revolution for gluers.
- Low mass to be moved and then a high value of the motor inertia to driven inertia ratio.

With this requirement the stepper motors are much better than the CC motors available at that time and probably even better than today's brushless motors. Another plus of the stepper motor is the better thermal behaviour with reduced heat dissipation.

Before starting the design, some tests were made, showing that a small 3.4 in. in diameter and 5 in. long motor, the so called 3450 size, is able to drive a directly coupled wheel with the same inertia as the motor, with good performance. In particular, the motor is capable of making one turn in about 54 ms with 60% of duty cycle.

On the basis of this test, it was decided to use direct coupling of the motors with driven parts without any gear reducer to simplify the design and reduce complexity, cost and noise source.

In order to reduce complexity, it was decided to unify motor sizes in all applications because the inertia and speed were similar for different unit. The stepper motor is typically not very precise because it has a low retention torque for small displacements but is very repetitive and, a strong point, is very well suited to digital control.

A drawback of these motors, at that time, was the presence of resonant torque oscillation at a given speed: we will see how this drawback was overcome.

Having selected the motor type and size, the next step is to look at the law of motion.

Almost all industrial motors application use an "algorithmic generated" motion law, variable during machine utilization: for example, in NC machines the law of motion is different for any "machining case", linear circular, spline, and is therefore generated at the running time. In the case of packaging machines, there is no need to operate in this way and the law of motion may be "frozen" (i.e. kept indefinitely), generating a real "electronic cam" that repeats the same movement continuously.

A polynomial was chosen for the possibility to match the motor capability and the functionality required by the packaging process: one revolution of the motor is made by 200 steps and the "cam" is defined by 200 points.

The use of such "frozen" law of motion also has a psychological benefit because it is easy to understand and evaluate by mechanical design, thus easing mechatronic integration.

The "table" of motion in Fig. 9.9 allows the definition of a law of motion very well optimized by looking at the torque generated by the motor and the actual acceleration of the system.

A program was developed on one of the first PCs available, for better control of the cam generation. With this program it is easy to define a law of motion in any way by an algorithm, and to obtain the stepping timing table. The program emulated a mechanical cam design program receiving in input the required law of motion and giving in output the "electronic cam program table".

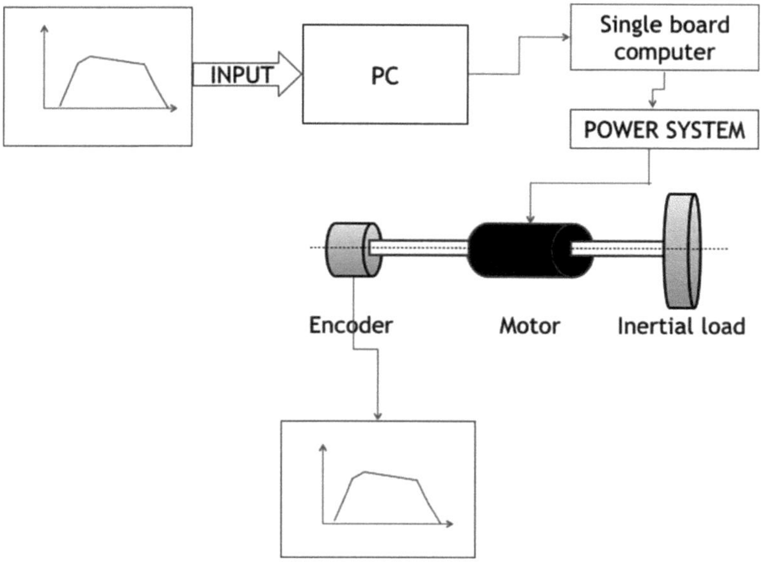

Fig. 10.15 System used to define and test the "frozen" law of motion of stepper motors

To speed the cam design and the law of motion, testing at a test bench is carried out connecting the output of the PC to the SBC (Single Board Computer) that will control the motor. An encoder is inserted on the shaft, in order to record the real behaviour of the motor to check the real law of motion. The layout of this innovative design tool is shown in Fig. 10.15: the ideal low of motion is given as input at a PC that computed the 200 line timing and then drives the stepping motor; a completely independent system recorded the law of motion in order to evaluate its quality.

With this testing system it is possible not only to design and test a given law of motion but also:

- To test the complete group consisting of motors and their driven parts.
- To optimize the law of motion looking at the real behaviour of the motor.
- Generate the final table to be installed on the machine as a read only memory.

The stepping motor behaviour is very particular and, being open loop controlled, is characterized by the "missing step" fault: when resistance is higher than the generated torque, the torque falls almost to zero and the motion control is completely lost. To avoid such crush the designer is able to modify the table point by point, taking into account the real slip between the rotating magnetic field and the rotor position. The careful design of this law makes it possible to obtain much higher performance from standard commercially available motors, than that imagined by stepping motor design. This plus is the origin of the long series of "problems to be solved" in the motor command chain and in the motor itself.

High performance means high current, thus high heat generation: two problems that arise when it is possible to test the motors under continuous testing at top cycle time. The power unit, commercially available, is capable of feeding around 10 A at 150 V but, at that time, the reliability of transistorized power units was bad and it took a long time to obtain a good one.

There was no commercially available system to develop this current and we were obliged to push an external producer to boost its unit to feed the current required: we found a small company with a willing "design-owner" that started the requirement improvements.

The next obstacle was generated by the overheating of the motors. Stepping motors, in general, have good thermal performance because much of the heat is generated by an eddy current inside the stator and in the winding that are all located on the stator. Generating optimal law of motion and the use of high current's drivers gave rise to a new problem: stepper motor overheating. In fact the law of motion required a very high current, at the given voltage, and motors were quickly showing a too high temperature.

Motors were designed and rated by old PWM, working at relatively low frequencies, while the last generation of PWM transistor based were working at a higher frequency that originated heat. But we had no choice and a complete water cool system was designed with a small chilling unit.

10.3 Control System

A layout of the control system is shown in Fig. 10.16. The absolute one turn encoder gives the phase advancement of the machine while the switching board gives the proper trigger to the computers. The triggering time may be easily modified by decadic switch and allows easy tuning of the machine phases [1].

A first consideration on general machine control system starts after the definition of the logical functionality of the whole machine that is characterized by:

- A core mechanical group.
- A series of functional sub-assemblies performing specific tasks such as:
  - Paper feeding.
  - Foil feeding.
  - Stamp feeding.

1. The first decision was to make functional independent groups, each one with its own control system plus a central supervising control unit for performing a second level control. The reasons are the following:

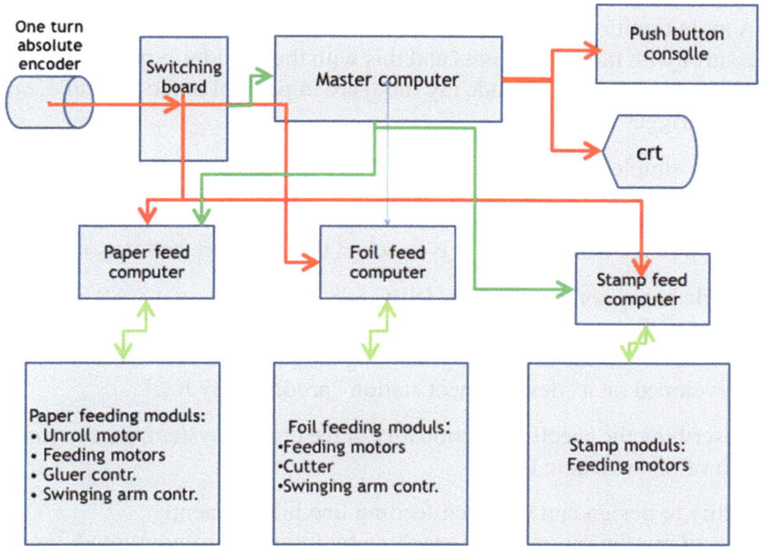

Fig. 10.16 Lay out of the control system: In red single wire lines, in green eight time parallel lines

- The goal of this strategic decision is to speed up the design, developing, mainly in parallel, the testing phases.
- Another advantage is the possibility to duplicate the control system for the paper and foil (that are rather similar).
- A further strong point of this modular solution is complexity reduction of the software to be developed and the multiple utilization of the same hardware modules.

2. The real time control outside the functional modules is brought to a minimum in order to reduce the criticality of the software that must be developed. Because timing is a very critical factor, each functional group receives a very simple "trigger" signal at each machine cycle and then it works autonomously without any further data interchange with the main central system.

    - An innovative synchronization system is developed that allows each functional group to receive precise timing through a very simple specific interface, in order to take any real time activity away from the central control unit.
    - The timing device consists of an absolute encoder connected with the main machine shaft (performing one revolution each machine cycle) and a hardware board producing trigger signals at a given rotation angle: the three triggers are then sent to the three control units to trigger the start up of their operating cycles.
    - These very simple systems take all the real time functionality to the machine level, letting the functional modules run in a complete autonomous way.

- A more traditional parallel, eight point to point lines to connect the peripheral control with the central units and this with the encoder, a push button console and a colour CRT (Cathode ray tube) are in parallel to these simple, one wire each, trigger linei.

This very simple, two layer layout and its connection are the "logic" design of the control system.

The next step in the design process is to select the hardware and its software.

- Intel single board computer model ISBC 8080 consisting of a 8080 8 bit CPU and completed with digital I/O line is selected.
- It is compulsory to use as the programming language the Intel assembly language to be developed on a "development station" produced by Intel.

Before describing the specific functionality of the control system, we can summarize the benefit of this strategic layout:

- Possibility to design and test each feeding line independently.
- Easiness of configuration of the machine by inserting different modules such as, for example, pre-cut blanks.
- Component modularity because the same SBC, single Board Computer, is used four times in the machine.
- Same software language for each component.

Moreover, the timing of the machine is not in any way linked with any hardware or software treatment, and, for this reason it gives rise to very clean and transparent running.

To evaluate the innovativeness of this layout, one has to compare this solution to the state of the art of the time of conceptual design, when only mechanical systems without independent motors were used, no independent functional unit, no computer control and no CRT display.

Let us look now at the main functionality of the different unit.

10.3.1 Peripheral Control Units

The three units are rather similar and so we describe the functionality of the control of one only, the foil feeder, that is intermediate in complexity with respect to the other two.

The function may be divided in three parts:

(a) Cycle feeding of foil.
(b) Swinging arm positioned and ribbon tension control.
(c) Presetting of the foil position, threading of the ribbon, loading unloaded coils.

10.3.1.1 (a) Cycle Feeding of Foil

The first function has already been described and is the main, synchronous function; it starts on arrival of the trigger from the encoder and is performed at standard speed. Because motors always follow the same law of motion, the execution time is always the same and corresponds to the machine top speed. During this activity, the CPU is devoted to this task only because the board must send "step" commands with high timing precision. At the machine start, the "V" shaped group must have at least one of the two coils pre-positioned. At each trigger signal, the stepper motors push forward in a position to mate with the arbor, the foil movement is always made at full speed and a series of photocells control its presence at the cycle start time and its absence at the cycle end.

A very important function is the switch from one coil to the other; the function may be triggered by different "emergency" conditions such as: ribbon break, lack of proper printing on the ribbon, jam, operator control. It was possible to perform this switch at full machine speed.

This functionality was present, at that time, only in the ALFA machine!

10.3.1.2 (b) Swinging Arm Positioning and Ribbon Tension Control

These series of functions guarantee a constant tension of the ribbon, obtained by the control of CC motors on the bobbin shaft and by the swinging of arm position. To avoid high acceleration of the coil, the swinging arm is kept in two positions:

1. Near its maximum swing, thus having the maximum quantity of ribbon stored when the ribbon is standing, ready to be used. In the case of start of the line, the ribbon storage gradually tends to empty and the coil is gently accelerated at steady state running speed.
2. Near its minimum swing, thus having the minimum quantity of ribbon stored, the line is feeding the machine. In this case of a sudden stop of the line, the stored ribbon quantity is increased, allowing a smooth slow down of the coil.

10.3.1.3 (c) Presetting the Foil Position, Threading of the Ribbon, Load Unload Coil

These miscellaneous functions are triggered by the push button console and allow human intervention for unloading empty coil, inserting a new one and threading the ribbon.

At the machine start, the "V" shaped group must have one at least of the two coils pre-positioned and then, at each trigger signal, the stepper motors push forward in a position to mate the foil movement with the arbor. It is always made at full speed and a series of photocells control its presence at the cycle start time, and its absence at the cycle end.

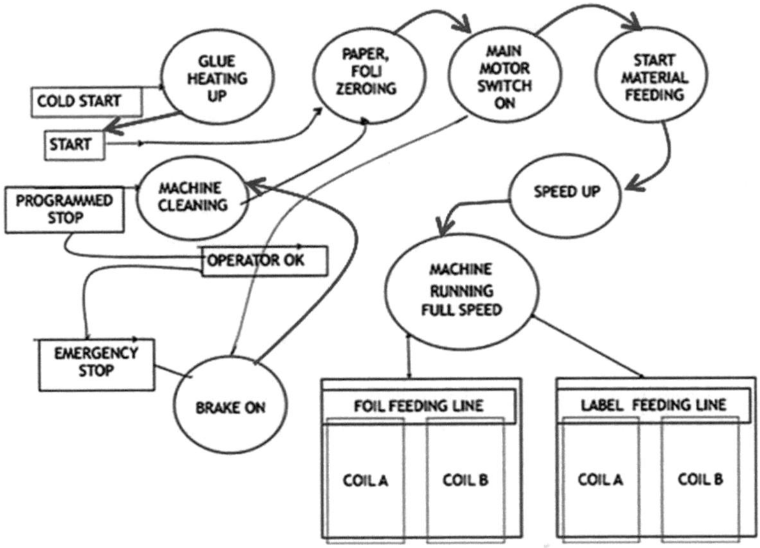

Fig. 10.17 A simplified diagram of main machine state (circle) and relative trigger command (box)

10.3.2 Central Unit

The main task of the central unit is the machine state control, without any strict real time constraint. The main operating states are shown in Fig. 10.17.

The central unit is in charge of managing the pack status through the machine with a sort of shift register simulating the real content of the material flowing in the packer. The register has a number of cells equal to the pack positions and the content of each cell gives information about the "quality" of the pack. The aim of the register is to manage the final ejection of bad quality cigarette packs, to stop and start timing of the wrapping material feeding. *This internal model of the machine is one of the most useful feature for the end user.*

The last functions of the central unit are:

- To supervise the two material feeding lines.
- To interface the CRT display and its push buttons.

10.3.3 Safety Circuitry

Operator safety and machine integrity control are carried out by traditional "fail-safe" type wired relay-based circuitry and in agreement with safety laws. The main components of safety circuitry are:

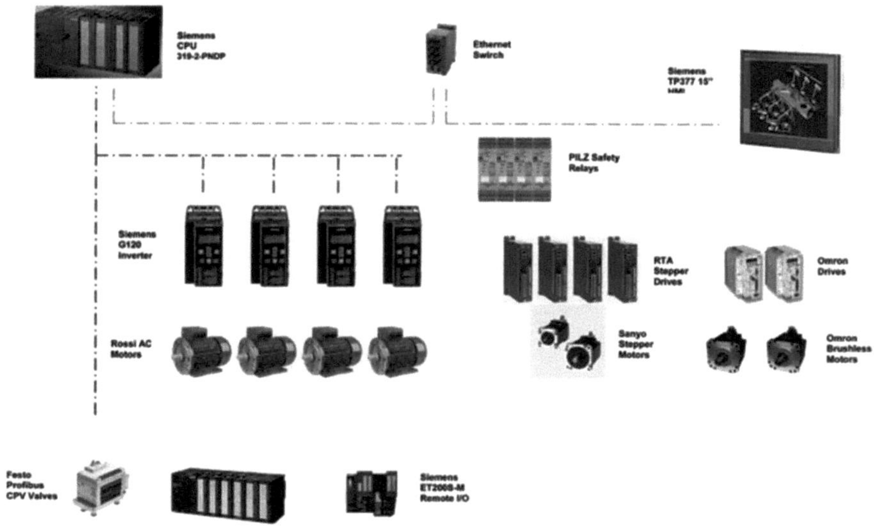

Fig. 10.18 The control system version based on a Siemens PLC in the 2017 production version

- Emergency "red mushroom" buttons connection.
- Safety guard closure.
- A series of "hit detector" to detect collision of mechanical parts of the machine, main belts breakage, etc.
- A series of component overheating detectors.

> Later on, the entire control system was completely modified, mainly by the non availability of the Intel SBC. A SIEMENS PLC as shown in Fig. 10.18 was used.

10.4 Sensors

A typical, unwanted behaviour of any packaging machinery is the frequent occurrence of "jamming" situations with production interruption without any damage to the machine itself.

Traditionally, jamming causes packaging material damage, materials spread around, presence of material in wrong positions, production of faulty packs: the "faulty mode" for any machine is almost infinite. Similarly, there is a very large list of jamming "immediate sources" such as: too much or insufficient glue, paper wrangling or tearing, feeding of two or more labels, paper incorrect position, etc.

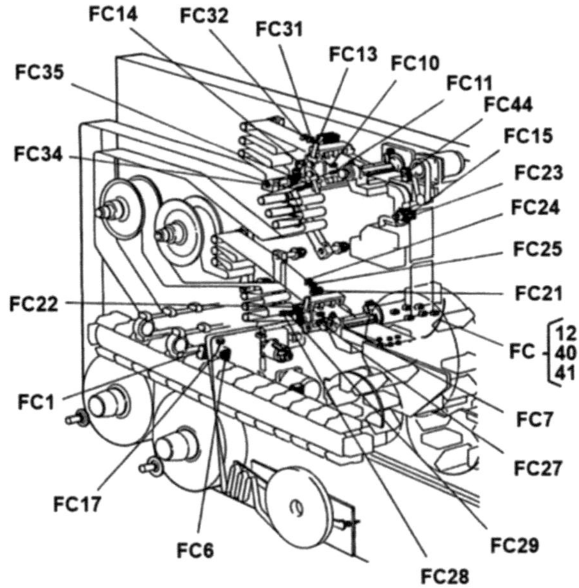

Fig. 10.19 Sensors, photocells, optic fiber sensors' micro-switches

Jamming always determines a machine stoppage automatically or triggered by a supervising person. Jamming reveals another two types of problems:

- It is difficult to spot where it is originated.
- It needs a manual "cleaning" operation.
- It requires a new start up of the machine, often difficult due to bad temperature setting, glue hardening, etc.

It leads to time consuming "cleaning" of the machine: in a packer running at 500 packs/min with a stopping time of only 0.5 s the machine is "buried" by 80 cigarettes spread around.

Last, but not least, the restart after any stoppage may be difficult, due to the fact that many parameters are affected by stoppage as: temperature, paper tension, gluing. In short: *jamming is always the crucial point in any packaging system*.

Usually, packers have a certain number of "sensors", such as micro switches and photocells that trigger the machine stoppage but, unfortunately, rarely indicate the "triggering behaviour".

For this reason the ALFA is fitted with a very large number of such devices, about 40 sensors placed as near as possible to the main "critical areas of the machine: the sensors" data are "interpreted" by each peripheral control system that is in charge of sending a message to the central unit and then to the CRT. Overall, there are almost 200 error messages, Fig. 10.19.

Table 10.1 Data on ALFA sensors system

Controlled entity	Control
• Paper roll	• Axial error
	• Conical Shape
• Paper thread from roll to arbor	• About a dozen of triggered photocells
• Foil thread from roll to arbor	•About a dozen of triggered photocells
• Cigarette feeding	• Cigarette presence
• Bad pack shape	• Skewness of label
• Folding station	• Presence/absence of mater. at a given time
• Pack appearance	• Print control warning print double check

The central system is also in charge of controlling special operation, such as:

- To switch from the running coil to another coil.
- To stop the machine in emergency or slow mode.
- To double check the feasible signal.
- To warn the operator.

The particularity of the sensor system was to read their status only at a given "time window" in order to detect any misrunning in real time.

The following table, albeit not complete, may give an idea of the ALFA sensor system, Table 10.1.

Runnabilty, as defined before, is good not only in a well managed company but also in problematic situations.

Many of the sensors perform a double check at each machine cycle in order to spot faulty sensors. Sensors are in mainly the fiber optic type to allow remote positioning and to facilitate repair.

10.5 Cost

It may be interesting to make some considerations about the production costs of the ALFA packer in spite of the years that have passed, the high inflation rate, the big technological improvements and, last but not least, the change of currency from Lira to Euro.

Moreover, it is well known that the cost of a product is a rather fuzzy value because it is linked to the rules used to calculate the overhead cost, jigs, mould and fixture cost, reference batch size, etc.

Forgetting "monetary" values, it may be interesting to look at the percentage proportions of the cost. The following considerations are made under the assumptions:

- To compute only labour direct costs without any overheads.
- Not to add tooling costs calculated for moulds, jigs, fixtures.

- To refer to a batch of ten productions, mainly made with NC machines.
- To refer to market component prices without any inventory costs.
- To compute raw material costs.

10.5.1 Cost Breakdown

From machine delivery time to machine running at full speed there is a relatively long period of time and considerable costs. In general, packaging machine are installed at the customer's plant by a crew of "external assembly specialists," i.e. a group of workers with complete knowledge of the machine who work to install and test the machine in its real working condition and to carry out the handover to the customer personnel.

Start up of the machine is not a trivial job, because one must tune the machine to operate with the actual material like paper, glue, cigarettes, ... and local personnel: in fact, the testing of the machine at the workshop is inevitably very limited due to the difficulties in having the proper quantity of tobacco and other materials. (At full speed the machine uses about 60 kg of tobacco per hour and produces 30,000 packs).

The external assemblers are a key element in the success story of a machine and it would be nice to write about their experience: we refer to [4] to have a general description of this type of activity.

In the ALFA case, we have no actual data but we may guess that at least three months were necessary to start up at full speed.

Referring to the time of production of the first batch of 10 machines we may have the following rough cost percentage breakdown ex works.

• Total direct cost (no overheads)	100%
• Design specific machined parts (around 1500 bluprints)	15%
• Commercial mechanical components	20%
• Commercial electronical components	45%
• Machine assembly	15%
• Raw materials	5%

The estimated, (no actual data available), production cost, (Lira 1984), was about 170 Million Lira that, deflated to Euro in 2017 is around 230,000 €.

It would be interesting to compare these data with the ones of previous traditional machines but there are no figures available. By memory, we may say that a 300 packs/min SASIB model IS 5000 had a cost around 30% of the ALFA and thus the cost per packs/min was around half of the new ALFA.

Development costs: at the time of the first delivered prototype was, deflated to today's Euro, about 10 million €, i.e. about 50 times the production cost.

Selling prices: information is not available and, moreover, is difficult to define because inter-company transfer and, in general, the selling price includes training, spare parts, guarantee, ... all data difficult to define, but we may assume a selling price around 1 million Euro today.

To complete the machine data records we may say that about 150 units were sold overall.

10.6 Additional Information

Before closing this chapter it may be interesting to provide same additional info that may be useful to better understand the whole picture.

10.6.1 Technology

During machine development, few technological aspects were attacked and, in general, materials and machining processes were the same as the previous SASIB products.

Almost all moving elements were made from aluminum alloys, machined from market available raw materials; large parts and all wheels were made from sand cast light alloy. The machine had very few parts broken during testing and in its useful industrial life. Some machined aluminum alloy parts were substituted with hot forged aluminum parts with good results.

Magnesium was tested but its use was discarded quickly.

Inside the design team there were no specific manufacturing experts but all technicians had a good knowledge of manufacturing processes and it was possible to quickly call specific meetings with production people.

We were attracted by carbon fiber technology: an alternating moving component was made by a specialized composite company [3]: it was wonderful in shape and rigidity but there was no time to complete the design. We tried to use carbon fiber rod plated with smooth chromium surface but without success.

The most demanding parts, the worm type pre-charged grinded index cams were purchased; the desmodromic cams were NC machined and surface flame-hardened.

10.6.2 Friendly Customer

At the project start, the preliminary design was shown to the main potential customers and, at an early stage, it was decided to bet on the innovation prone management of the Philips Morris company because of the strong concentration of production facilities at

Richmond, where the SASIB Company of America, a fully owned service company, was located.

In this light, PM technicians were continuously invited to look at the design progress and the first prototype was installed at the PM Richmond plant. The bet was a winner, because PM ordered the first batch of a dozen machines. In this way we quickly obtained the "reference" that was so necessary for a lunch of a completely new innovative machine.

10.6.3 Prototype Department at SASIB

Before starting the project a new laboratory was set up for prototype assembly and testing with the following activity:

- To manage prototype and sub-assembly, assembly and testing.
- To acquire commercial components and machine parts.
- To perform noise and vibration analysis by high speed filming.

The lab was working most of its time for the ALFA project in close cooperation with design services; the head of the lab had the full responsibility and freedom to buy any components without going through the standard company bureaucratic authorization procedure, speeding up prototype completion.

The measurement activity was also very important and its analyses were fundamental for spotting faults, to measure and reduce noise emission.

To identify noise source and critical vibrations, clearances and shocks, a very sophisticated procedure was based on filtering noise in harmonics and for the machine cycle phase: in this way it was rather easy to spot the origin of any specific noise. This method was presented at a North Caroline conference but was not published due to lack of internal resources.

10.7 What We Have Learned from an Automatic Machine Design

The design case story is instructive from various points of view. In particular, it shows that the industrial environment has its own rules and limits that are different from those of an experimental laboratory.

First of all, technical tradition is very important, represented by the specific company know-how and by human behaviour that connects different responsibilities. Even the comments have the same importance as the main story.

Secondly, it must be accepted that the classic methods of design, based on theoretical problem-solving thought, must come to a compromise with the reality that often shortens certain phases, prolongs others and places new constraints on the project.

In any case, a certain evolutionary thought (one thing comes from another thing) that implies that a design always derives from a previous one, often dominates the industrial environment, with a mix of advantages and disadvantages.

References

1. AMMA (1983) Aziende Meccaniche Meccatroniche Associate. In: EDA-Torino (ed). https://www.amma.it/, Via Vela, 17 - 10128 Torino
2. Bleisch G, Majschak JP, Weiss U (2010) Verpackungstechnische Prozesse. Behr's Verlag Hamburg
3. Dallara (1985). https://www.dallara.it/
4. Levi P (2014) La chiave a stella, vol 1. Einaudi Inc., EAN, p 9788806219529
5. Meneghetti U (1984) Cam profile calculations for two different laws of motion. Mechanics, Bologna University, Unpublished Consultancy, Dept

Appendix
Packaging Valley History

A.1 Peculiarity and Size

Any industrial product is made by its design team but is strongly influenced by the local social environment, as in the Silicon Valley paradigmatic case.

In the present case, another "valley" strongly influenced the people involved and the technologies employed: the *Packaging Valley*, an area of a few hundred $sq.kms$ around the Italian city of Bologna.

The packaging valley phenomena is rather typical of the Italian economic activity. The Packaging valley is characterized by the percentage (about 20%) of packing machine producers over the worldwide market of this product. The Packaging valley has a large overlap with another important "valley" of the motor vehicles producers in the Emilia-Romagna Region, such as Ferrari, Maserati, Lamborghini, Ducati, Dallara, Toro Rosso, etc. The two valleys are geographically overlapped, share part of their technicians and have a large interchange of ideas and knowledge. The district phenomenon is not peculiar to Italy only but its behaviour is much stronger here than in any other country worldwide [1–3].

Any Italian industrial district has its peculiarity in origin and shape and often it is difficult to recognize its roots and the reason behind its development: in many case there is no "natural" resource peculiarity such as water or a local mineral. In many cases the main driving force is typically cultural, i.e. is connected with intangible local values.

In the present case of the packaging valley, boundaries are usually made coincident with the two "provinces" of Bologna and Modena, regarding design, production, sale, and after service activities of locally produced machinery.

In this globalization time it is difficult to define business data relative to a given territory: many local industries have own subsidiaries located worldwide or their foreign parent company still located abroad in their original location; on the contrary many local industries have been acquired by foreign companies.

A. Freddi and M. Salmon, *Design Principles and Methodologies*, Springer Tracts in Mechanical Engineering, https://doi.org/10.1007/978-3-319-95342-7

Packaging Valley is characterized by the following data:

- Italian turnover (year 2016): 6.6 million €, equal to 18% of world turnover.
- Percentage of exports from Italy: 80% (average 2012–2016).
- Percent turnover of packaging machine manufacturing in Emilia-Romagna over total manufacturing in Emilia Romagna: 62%.
- Percent turnover of Emilia Romagna over total Italian turnover: 62%.
- Packaging companies in Emilia Romagna: 222.

Another characteristic of the packaging machine market is a rather strong concentration in each market segment. In many sectors, mainly for the richer or larger ones, often two producers together hold 50% of the market share. Such a strong position, the large number of patents, the technical difficulties of each product and the small dimensions of each market niche are at the basis of a rather mature market with very few newcomers. The exception is when there is a new product to be packaged such as floppy disks, tape cassettes for video or audio recording, cigarette replacement cartridges, coffee capsules, etc.

Cigarette packaging is a very large and rich sector characterized by few large companies: at the time of this *case study*, around 1980, the cigarette packaging machine producers were two companies with almost the entire market share:

- GD, Italy, a Bologna based company, producing soft and hard box machinery having by far the largest market share.
- SASIB, Bologna based too, producing soft pack machines.

Plus a few other producers with a very small market share:

- Schermund, Germany, producing soft pack machinery.
- Focke, Germany, producing hard box machinery.
- Molins, UK, producing only hard box machinery.

At that time, the market was dominated by the GD machines at 360 packages/min, followed by SASIB with a 300 packages/min machine.

A.2 Historical Background

The reasons for such an exceptional concentration of similar industries are numerous and very different in type and timing, in short the main driving forces were:

- In the 15th century: Bologna was European leader in silk production and, at that time, many special purpose machines were built locally [4]. The local *Industrial Heritage Museum* (*Museo del patrimonio industriale*) has interesting replica of large complex of silk "torcitoi" or "silk twisting machines".
- Four centuries later, the Bologna Region became an agricultural based economy, mainly devoted to the "rich" segment like fruit and vegetable products, developing and utilizing many "clever" agricultural machines.

- A third driving force was a very advanced Technical School the "Istituto Aldini Valeriani".

The recent situation triggered, in the early 1930s, a very peculiar "chain reaction" where a very "first case" of automatic packaging machines gave rise to a large series of spin-off companies, all located in a very concentrated district.

(The seminal case was the automatic machine, produced by the ACMA company in 1924 to fill and seal powder inside a sachet to prepare, at home, sparkling water). Many skilled technicians, in the *WWII* period, left ACMA and founded a number of small new companies.

Around 1980 there were many companies operating in the packaging sectors, everyone strongly devoted to a particular "niche" of the market. This chain process went on until the end of the 20th century, when an inverse process occurred and a few large companies started an acquisition process, leading to a concentration of the market in a dozen companies. The concentration process was further accelerated by the presence of companies and investments coming from abroad, as in the case of the Tetra-pack Group.

To complete this essential packaging valley description we may also say that for many years the companies were organized in a very vertical way with large internal workshops for part machining and a very limited use of external companies. When the expansion rate became high, many companies changed their business strategy and made use of external suppliers for part machinery, cabling and sophisticated electronics control systems.

Today, the "district" is made up of a few "OEM (Original Equipment Manufacturer)" producers, many small workshops, and a series of "sellers" of electronic devices with local engineering capacity. Some of the worldwide leaders in electronic component production have research ("excellence") centers just in Bologna to serve the local industries and to develop packaging oriented components.

A.2.1 Social Values

One of the particularities of the valley is, no doubt, the strong personal link and the "competition" aspect of the design activity. The first aspect is characterized by the fact the large majority of the persons working in the packaging sector were born locally and went to the same school: Technical School and/or University and, moreover, they share sports activity, football team supporters, clubs, etc.

On the contrary, the strong competition between companies limits any leakage and avoids any interchange of people. Nevertheless, a certain amount of insight activity, inspection "spying" of competitors at customer plants is always present and all companies are very active in patenting every possible invention.

A.2.2 SASIB History

The SASIB Company, as a general purpose mechanical workshop, was founded in 1915 by Scipione Innocenti, Fig. A.1 a workman from a humble family of Bologna.

The first products of the company were railway level crossing barriers and, later, signalling and rail points.

Around 1937, when importing any American machinery was forbidden, SASIB started the production of a cigarette packaging machinery under licence from the American company AMF that, later after the war, acquired the control of the SASIB company.

From 1937 up to early 1970, SASIB became the leader in soft pack cigarette packaging, exporting its product into almost every country. The SASIB main product was a soft pack automatic packaging machinery, keeping the original layout and speeding up the packaging from 125 pack/min to 300 pack/min in the last years of the 1970s.

The American company AMF, owner of SASIB, then sold the company because the ownership of a tobacco company was not in line with the "green" image of manufacturing of amusement products such as golf clubs, bowling systems and alike.

At the beginning of 1978, Mr. Carlo De Benedetti bought the SASIB Company and later Olivetti. So it was easy to transfer know-how from Olivetti to SASIB.

While AMF ran the company in a very conservative manner, under De Benedetti control the company was revitalized with new organization and considerable investment in R&D. The integration of traditional SASIB mechanical know-how with modern Olivetti know-how was a key factor for the design of the ALFA packer.

At the ALFA project start up, SASIB consisted of two business units for tobacco machinery and railway safety equipment. The overall characteristics of the company were:

- Total turnover (1984): 110,000 million Lira.
- Tobacco turnover (1984): 60,000 million Lira.
- Tobacco export: 98%.
- Total personnel: 1200.
- Tobacco BU R&D personnel around 100.

The company was located in one central plant in Bologna, with a fully owned subsidiary: SASIB Co. of America, in Richmond VA. (USA).

At the end of 1982, the SASIB CEO, Frisoni, died and the CEO role was given to Mr. Giancarlo Vaccari, Fig. A.1 and, at the same time, Mr. Franco Debenedetti, brother of Carlo De Benedetti, became President of SASIB. A clever inventor, a collaborator of one of the authors, was Mr. Elio Pagella, Fig. A.1.

From 1980, the strategy of the company was completely revised: the company was quoted on the Milan stock exchange in 1984, and became a conglomerate holding, buying many other industries in the food machinery section. Later the company was sold in 2011, to the Coesia group, owner of the historical competitor GD.

(a) *Scipione Innocenti: the Sasib Founder*

(b) *From left: Mario Salmon: (R&D Director in Sasib at the time of Alfa design), Franco Debenedetti, (SASIB President at that time)*

(c) *Elio Pagella: Consultant and Chief of mechanical design*

(d) *From left: unknown, Rodolfo De Benedetti, Giancarlo Vaccari (CEO after 1982), unknown, at the the prototype*

Fig. A.1 Some pictures of the Sasib founder and other people involved in ALFA design

When COESIA became the new owner, large parts of the original SASIB crew had left the company and the relative know-how had left.

COESIA did a wonderful job of, say, reverse engineering, writing new detailed documentation of the BOM, the assembly instruction and the user manual and refurbishing all drawings in CAD 3D: all the pictures of last two chapters are extracted from this documentation, thanks to the permission given by Mr. Lena the current, (2017), CEO of SASIB.

References

1. Breschi S, Malerba F (2005) Clusters, networks and innovation. OUP Oxford
2. Fortis M, Carminati M (2015) The automatic packaging machinery sector in Italy and Germany. Springer Briefs in Business. ISBN: 978-3-319-12763-7
3. Lipparini A (1995) Imprese, relazione tra imprese e posizionamento competitivo. Etas Libri. ISBN: 88-453-0779-4
4. Several Authors (2018) Cronologia a Bologna dal 1796 a oggi. http://www.bibliotecasalaborsa.it/cronologia/bologna/1918/102

Index

A. Freddi and M. Salmon, *Design Principles and Methodologies*, Springer Tracts in Mechanical Engineering, https://doi.org/10.1007/978-3-319-95342-7

MIX
Papier aus verantwortungsvollen Quellen
Paper from responsible sources
FSC® C105338

If you have any concerns about our products,
you can contact us on
ProductSafety@springernature.com

In case Publisher is established outside the EU,
the EU authorized representative is:
Springer Nature Customer Service Center GmbH
Europaplatz 3, 69115 Heidelberg, Germany

Printed by Libri Plureos GmbH
in Hamburg, Germany